LECTURES ON COMPLEX ANALYTIC VARIETIES:
THE LOCAL PARAMETRIZATION THEOREM

BY

R. C. GUNNING

PRINCETON UNIVERSITY PRESS

AND THE

UNIVERSITY OF TOKYO PRESS

PRINCETON, NEW JERSEY
1970

L.C. Card: 73-132628

I.S.B.N.: 0-691-08029-1

A.M.S. 1968: 3244

Published in Japan exclusively by the University of Tokyo Press; in other parts of the world by Princeton University Press

Printed in the United States of America

PREFACE

In introductory courses on complex analytic varieties, it is customary to begin the local description of irreducible subvarieties by choosing a system of coordinates in the ambient space $\mathbb{C}^n$ such that the subvariety is in a particularly convenient position, for example, such that the subvariety appears as a branched covering space of a coordinate hyperplane $z_{k+1} = \cdots = z_n = 0$ under the natural projection mapping. The existence of such coordinate systems, together with a catalog of the elementary properties of analytic subvarieties in terms of these coordinate systems, comprise what may be called the local parametrization theorem for complex analytic subvarieties. Once this has been established, it is relatively easy to derive the standard local properties of analytic subvarieties, and the way is then clear to proceed to more advanced topics, either on the local or the global level.

These lecture notes treat the local parametrization theorem, assuming some background knowledge of the general function theory of several complex variables. They contain the material common to the first parts of several courses of lectures on complex analytic varieties that I have given in the past few years. They go further in various directions into the properties of complex analytic varieties than some recent texts on the subject (such as L. Hörmander, An Introduction to Complex Analysis in Several

Variables; or R. C. Gunning and H. Rossi, Analytic Functions of Several Complex Variables), and have a rather different point of view and emphasis than other texts (such as M. Hervé, Several Complex Variables, Local Theory; R. Narasimhan, Introduction to the Theory of Analytic Spaces; or S. Abhyankar, Local Analytic Geometry). Of course, every author feels that his own organization of the material is in some ways superior to that currently available in the literature.

The first section is a survey of prerequisites from the general function theory of several complex variables. The second and third sections cover the local parametrization theorem for complex analytic subvarieties of the space of several complex variables, and some of its immediate consequences. The fourth section introduces the notion of an analytic variety (also known as an analytic space) as an equivalence class of analytic subvarieties, abstracting those properties of analytic subvarieties that can be considered as being less dependent on the particular imbedding in the space of several complex variables; there seem to be definite didactical advantages to stressing this distinction between varieties and subvarieties. The fifth and sixth sections cover those aspects of the local parametrization theorem that remain meaningful for analytic varieties; the fifth section treats branched analytic coverings, which correspond to the projections of analytic subvarieties on coordinate hyperplanes, and the sixth section treats simple analytic mappings, which correspond to partial projections in the local parametrization theorem for complex analytic subvarieties.

I should like to express my thanks here to the students who have attended the various courses on which these notes are based, for all their helpful comments and suggestions, and to Elizabeth Epstein, for her customary beautiful job of typing.

<u>A remark on the notation.</u> The usual mathematical notations are used throughout, except that $\subseteq$ is used to denote general set inclusion while $\subset$ is used to denote proper inclusion (excluding equality). There is no separate notation used to distinguish equivalence classes from representatives of the equivalence classes, in discussing varieties or germs of functions or sets; the additional notation is more burdensome and confusing than the systematic confusion of no notation.

CONTENTS

§1. A background survey

(a) Some familiarity with the local properties of complex analytic functions of several complex variables will be assumed from the beginning. The reader acquainted with the material contained in Chapter I (sections A through D) and Chapter II (sections A through D) of Gunning and Rossi, Analytic Functions of Several Complex Variables (Prentice-Hall, 1965), or in Chapter VI of Hörmander, An Introduction to Complex Analysis in Several Variables (Van Nostrand, 1966), will certainly have an adequate background. In order to establish notation, terminology, and references, a brief introductory review of this prerequisite material will be included here.

The ring of germs of holomorphic functions of n complex variables at a point $a = (a_1,\ldots,a_n) \in \mathbb{C}^n$ will be denoted by ${}_n\mathcal{O}_a$, or by $\mathcal{O}_a$ for short when the dimension n is either understood or irrelevant. This can be identified with the ring of convergent complex power series $\mathbb{C}\{z_1-a_1,\ldots,z_n-a_n\}$, by viewing $z_1,\ldots,z_n$ as the coordinate functions in $\mathbb{C}^n$. The ring ${}_n\mathcal{O}_0$ at the origin $0 = (0,\ldots,0) \in \mathbb{C}^n$ will also be denoted simply by ${}_n\mathcal{O}$ or $\mathcal{O}$, when there is no danger of confusion. The rings ${}_n\mathcal{O}_a$ for various points $a \in \mathbb{C}^n$ are canonically isomorphic in the obvious manner. Indeed, any nonsingular complex analytic homeomorphism from an open neighborhood of a point $a \in \mathbb{C}^n$ onto an open neighborhood of a point $b \in \mathbb{C}^n$ induces an isomorphism between the rings ${}_n\mathcal{O}_a$ and

${}_n\mathcal{O}_b$; the simplest such homeomorphism, which will be used for the canonical isomorphism, is just translation. Thus for studying the local properties of analytic functions of several complex variables, it is generally sufficient to consider merely the ring ${}_n\mathcal{O} = {}_n\mathcal{O}_0$. An arbitrary complex analytic local homeomorphism preserving the origin can then be used for further simplification as needed.

Recall that ${}_n\mathcal{O}$ is an integral domain; the product of two non-zero elements of the ring ${}_n\mathcal{O}$ cannot be the zero element, the germ of the function identically zero. Hence ${}_n\mathcal{O}$ has a well defined quotient field ${}_n\mathcal{M}$, the field of germs of meromorphic functions at the origin in $\mathbb{C}^n$. The ring ${}_n\mathcal{O}$ is also Noetherian; every ideal in the ring has a finite basis. The units of the ring ${}_n\mathcal{O}$, the elements of ${}_n\mathcal{O}$ having inverses in ${}_n\mathcal{O}$, are precisely the germs of functions which are not zero at the origin. Consequently the non-units form an ideal ${}_n\mathcal{W}$ in the ring ${}_n\mathcal{O}$, the ideal of all germs of functions which are zero at the origin; a Noetherian ring with this property is known as a local ring. The ideal ${}_n\mathcal{W}$ of non-units is clearly the unique maximal ideal of the ring ${}_n\mathcal{O}$; the residue class field ${}_n\mathcal{O}/{}_n\mathcal{W}$ is evidently isomorphic to the field $\mathbb{C}$ of complex numbers. Finally, the ring ${}_n\mathcal{O}$ is a unique factorization domain, an integral domain in which every non-unit can be written, uniquely up to the order of the factors and the units in the ring, as a finite product of irreducible elements; an irreducible element is one which cannot be written as a product of two non-units.

Two very useful tools in deriving these and other properties of the local rings ${}_n\mathcal{O}$ are the Weierstrass preparation and division theorems. Note that the ring ${}_{n-1}\mathcal{O}$ is canonically imbedded as the subring of ${}_n\mathcal{O}$ consisting of germs of functions which are independent of the last variable, corresponding to the imbedding $\mathbf{C}\{z_1,\ldots,z_{n-1}\} \subset \mathbf{C}\{z_1,\ldots,z_{n-1},z_n\}$. Between these two rings lies the ring ${}_{n-1}\mathcal{O}[z_n]$ of polynomials in the variable z_n with coefficients from the ring ${}_{n-1}\mathcal{O}$. The theorems of Weierstrass facilitate the natural induction step from ${}_{n-1}\mathcal{O}$ to ${}_n\mathcal{O}$ through the intermediate ring ${}_{n-1}\mathcal{O}[z_n]$. In more detail, an element $f \in {}_n\mathcal{O}$ is said to be <u>regular in</u> z_n if the germ $f(0,\ldots,0,z_n)$ is not identically zero, as the germ of a holomorphic function of the single complex variable z_n; the element f is said to be regular in z_n <u>of order</u> k if the germ $f(0,\ldots,0,z_n)$ has a zero of order k at the origin in the plane of the complex variable z_n. Given any finite set of elements $f_i \in {}_n\mathcal{O}$, there is a nonsingular linear change of coordinates in $\mathbf{C}^n$ making all of these elements regular in z_n. A <u>Weierstrass polynomial</u> of degree k in z_n is an element $p \in {}_{n-1}\mathcal{O}[z_n]$ of the form

$$p = z_n^k + a_1 z_n^{k-1} + \ldots + a_{k-1} z_n + a_k ,$$

where the coefficients $a_i \in {}_{n-1}\mathcal{O}$ are non-units. It is evident that a Weierstrass polynomial of degree k in z_n is an element of ${}_n\mathcal{O}$ which is regular in z_n of order k. The <u>Weierstrass preparation theorem</u> asserts that a Weierstrass polynomial of degree k in z_n is the generic form of an element $f \in {}_n\mathcal{O}$ which is regular in z_n

of order k, in the sense that whenever $f \in {}_n\mathcal{O}$ is regular in z_n of order k there is a unique Weierstrass polynomial $p \in {}_{n-1}\mathcal{O}[z_n]$ of degree k in z_n such that $f = up$ for some unit $u \in {}_n\mathcal{O}$. The <u>Weierstrass division theorem</u> asserts that if $p \in {}_{n-1}\mathcal{O}[z_n]$ is a Weierstrass polynomial of degree k in z_n, then any element $f \in {}_n\mathcal{O}$ can be written uniquely in the form $f = pq + r$, where $q \in {}_n\mathcal{O}$ and $r \in {}_{n-1}\mathcal{O}[z_n]$ is a polynomial in z_n of degree less than k; moreover, if $f \in {}_{n-1}\mathcal{O}[z_n]$, then necessarily $q \in {}_{n-1}\mathcal{O}[z_n]$.

The Weierstrass theorems are not really limited in applicability merely to the local rings; they can easily be extended to global situations, provided some care is taken with the domains of existence of the functions involved. For any open set $U \subseteq \mathbb{C}^n$ let $\mathcal{O}_U$ denote the ring of functions holomorphic in U. Suppose that U is a connected open neighborhood of the origin in $\mathbb{C}^n$, and that U can be written as the product $U = U' \times U''$ of an open set $U' \subseteq \mathbb{C}^{n-1}$ and an open set $U'' \subseteq \mathbb{C}$, the complex plane of the variable z_n. If $p \in \mathcal{O}_U$ defines a Weierstrass polynomial in the local ring ${}_n\mathcal{O}_0$, then p has the form

$$p = z_n^k + a_1 z_n^{k-1} + \ldots + a_{k-1} z_n + a_k$$

where the coefficients $a_i \in \mathcal{O}_{U'}$. Note that for each fixed point $z' = (z_1, \ldots, z_{n-1}) \in U'$ the polynomial $p(z_1, \ldots, z_{n-1}, z_n)$ in z_n has k roots; assume that all of these roots are contained in the open set U''. Under these hypotheses, the <u>extended Weierstrass division theorem</u> asserts that any function $f \in \mathcal{O}_U$ can be written uniquely in the form $f = pq + r$, where $q \in \mathcal{O}_U$ and $r \in \mathcal{O}_{U'}[z_n]$ is a polynomial in z_n of degree less than k. One rather immediate corollary of the extended Weierstrass division theorem is the following. Suppose that U is an open neighborhood of the origin in $\mathbb{C}^n$, and that $f_1, \ldots, f_k \in \mathcal{O}_U$ are functions whose germs at the origin generate an ideal $\mathfrak{A} \subseteq {}_n\mathcal{O}_0$; then there is a subneighborhood $U_0 \subseteq U$ of the origin in $\mathbb{C}^n$ such that any function $f \in \mathcal{O}_{U_0}$ whose germ at the origin is contained in the ideal $\mathfrak{A}$ can be written in the form $f = g_1 f_1 + \ldots + g_k f_k$ for some functions $g_i \in \mathcal{O}_{U_0}$.

(b) The use of sheaves proves to be a very helpful notational and organizational convenience in several complex variables. This is perhaps not really apparent at the outset, since notational complications are minimal and the interest is mostly in purely local phenomena. Consequently, throughout most of the present lectures sheaves will appear only incidentally as an alternative notation. Eventually, however, sheaves will be freely used; and some of the deeper semi-local properties of holomorphic functions, which are most conveniently stated in the language of sheaves, will play an important role. The reader should thus have some familiarity with analytic sheaves; an acquaintance with the material contained in Chapter IV (sections A through C) of Gunning and Rossi, Analytic Functions of Several Complex Variables, or in section 7.1 of Hörmander, An Introduction to Complex Analysis in Several Variables, will provide an adequate background. Again a brief review of this material will be included here, primarily to establish notation.

The sheaf of germs of holomorphic functions of n complex variables will be denoted by ${}_n\mathcal{O}$, or merely by $\mathcal{O}$ when no confusion is likely to arise. For any open set $U \subseteq \mathbb{C}^n$ there is a natural identification of the ring $\Gamma(U, \mathcal{O})$ of sections of the sheaf $\mathcal{O}$ over U with the ring $\mathcal{O}_U$ of holomorphic functions in U ; and for any point $a \in \mathbb{C}^n$ there is a natural identification of the stalk ${}_n\mathcal{O}_a$ of the sheaf $\mathcal{O}$ over the point a with the ring ${}_n\mathcal{O}_a$ of germs of holomorphic functions at that point. An <u>analytic sheaf</u> over an open set $U \subseteq \mathbb{C}^n$ is a sheaf of modules over the restriction ${}_n\mathcal{O}|U$ of the sheaf of rings ${}_n\mathcal{O}$ to the set U ; perhaps the simplest example is the <u>free analytic sheaf</u> of rank r over U , the direct sum $({}_n\mathcal{O}|U)^r = ({}_n\mathcal{O}|U) \oplus \dots \oplus ({}_n\mathcal{O}|U)$ of r copies of the sheaf ${}_n\mathcal{O}|U$. An analytic sheaf $\mathcal{S}$ over U is said to be <u>finitely generated</u> over U if there are finitely many sections of $\mathcal{S}$ over U which generate the stalk $\mathcal{S}_a$ as an ${}_n\mathcal{O}_a$-module at each point $a \in U$, or equivalently, if there is an exact sequence of analytic sheaves of the form

$$({}_n\mathcal{O}|U)^r \longrightarrow \mathcal{S} \longrightarrow 0$$

for some r . An important and often used semi-local property of holomorphic functions of several complex variables is given by <u>Oka's theorem</u>: for any analytic sheaf homomorphism $\varphi: ({}_n\mathcal{O}|U)^r \rightarrow ({}_n\mathcal{O}|U)^s$ over an open set $U \subseteq \mathbb{C}^n$, the kernel of φ is a finitely generated analytic sheaf in an open neighborhood of each point of U .

An analytic sheaf $\mathcal{S}$ over an open set $U \subseteq \mathbf{C}^n$ is said to be coherent if in some open neighborhood $U_a \subseteq U$ of each point $a \in U$ there is an exact sequence of analytic sheaves of the form

$$({}_n\mathcal{O}|U_a)^r \longrightarrow ({}_n\mathcal{O}|U_a)^s \longrightarrow (\mathcal{S}|U_a) \longrightarrow 0$$

for some r, s. It then follows from Oka's theorem that coherence is preserved under many standard algebraic operations on sheaves; for example, for any analytic sheaf homomorphism $\varphi: ({}_n\mathcal{O}|U)^r \longrightarrow ({}_n\mathcal{O}|U)^s$, the kernel and image of the homomorphism are coherent analytic sheaves, and for any exact sequence of analytic sheaves of the form

$$0 \longrightarrow \mathcal{R} \longrightarrow \mathcal{S} \longrightarrow \mathcal{T} \longrightarrow 0 ,$$

if any two of the sheaves are coherent so is the third.

§2. The local parametrization theorem for analytic subvarieties

(a) An analytic subvariety of an open set $U \subseteq \mathbb{C}^n$ is a subset of U which in some open neighborhood of each point of U is the set of common zeros of a finite number of functions defined and holomorphic in that neighborhood. Note that an analytic subvariety of U is necessarily a relatively closed subset of U. The subject of the present lectures is the local nature of such analytic subvarieties in the neighborhood of some fixed point of $\mathbb{C}^n$, which for convenience will be taken to be the origin. To make this precise, consider the set of pairs (V_α, U_α), where U_α is an open neighborhood of the origin in $\mathbb{C}^n$ and V_α is an analytic subvariety of U_α. Two such pairs (V_1, U_1) and (V_2, U_2) will be called equivalent if there is an open neighborhood $W \subseteq U_1 \cap U_2$ of the origin such that $W \cap V_1 = W \cap V_2$; it is readily seen that this is indeed a proper equivalence relation. An equivalence class of these pairs is called the germ of an analytic subvariety at the origin in $\mathbb{C}^n$; and these equivalence classes are really the subject of the lectures. Any germ can be represented by an analytic subvariety V of some open neighborhood U of the origin; but the only properties to be considered here are those that are independent of the choice of representative subvariety of the germ. In the notation and subsequent discussion there will be no systematic distinction between germs and representative varieties whenever there is no serious likelihood of confusion.

To each germ V of an analytic subvariety at the origin in $\mathbb{C}^n$ there is canonically associated an ideal in the local ring ${}_n\mathcal{O}_0$ called the ideal of the subvariety V at the origin and denoted by $\mathrm{id}(V)$, defined as follows:

$$\mathrm{id}(V) = \{f \in {}_n\mathcal{O}_0 \mid \text{there exist an analytic subvariety } V \text{ of an open set } U \subseteq \mathbb{C}^n \text{ representing the germ } V \text{ and an analytic function } f \in \mathcal{O}_U \text{ representing the germ } f, \text{ such that } f|V \equiv 0 .\}$$

It is clear that this is a well defined ideal in ${}_n\mathcal{O}_0$. In the other direction, to each ideal $\mathfrak{A} \subseteq {}_n\mathcal{O}_0$ there is canonically associated a germ of an analytic subvariety at the origin in $\mathbb{C}^n$, called the locus of the ideal $\mathfrak{A}$ and denoted by $\mathrm{loc}(\mathfrak{A})$, defined as follows:

$$\mathrm{loc}(\mathfrak{A}) = \text{germ represented by the analytic subvariety } V = \{z \in U | f_1(z) = \ldots = f_r(z) = 0\} \text{ of the open set } U \subseteq \mathbb{C}^n, \text{ where } f_i \in \mathcal{O}_U \text{ are analytic functions in } U \text{ whose germs in } {}_n\mathcal{O}_0 \text{ generate the ideal } \mathfrak{A} .$$

It is clear that this is a well defined germ of an analytic subvariety at the origin in $\mathbb{C}^n$; recall that any ideal $\mathfrak{A} \subseteq {}_n\mathcal{O}_0$ is finitely generated, since the local ring ${}_n\mathcal{O}_0$ is Noetherian. These correspondences permit a very useful and interesting interplay to develop between the geometrical properties of germs of analytic subvarieties at the origin in $\mathbb{C}^n$ and the algebraic properties of ideals in the local ring ${}_n\mathcal{O}_0$.

Several quite simple properties of these correspondences follow almost immediately from the preceding definitions; the proofs will be left as exercises for the reader. If V, V_1, V_2 are germs of analytic subvarieties at the origin in $\mathbf{C}^n$, and $\mathfrak{A}, \mathfrak{A}_1, \mathfrak{A}_2$ are ideals in the local ring ${}_n\mathcal{O}_0$, then:

(i) $V_1 \subseteq V_2 \implies \text{id } V_1 \supseteq \text{id } V_2$;

(ii) $\mathfrak{A}_1 \subseteq \mathfrak{A}_2 \implies \text{loc } \mathfrak{A}_1 \supseteq \text{loc } \mathfrak{A}_2$;

(iii) $V = \text{loc id } V$;

(iv) $\mathfrak{A} \subseteq \text{id loc } \mathfrak{A}$, but equality does not necessarily hold;

(v) $V_1 = V_2 \iff \text{id } V_1 = \text{id } V_2$.

Note that $V_1 \cup V_2$ and $V_1 \cap V_2$ are also germs of analytic subvarieties at the origin in $\mathbf{C}^n$, where the unions and intersections of germs are defined respectively as the germs of the unions and intersections of representative subvarieties. Indeed, if $V_1 = \text{loc } \mathfrak{A}_1$ and $V_2 = \text{loc } \mathfrak{A}_2$, then

(vi) $V_1 \cap V_2 = \text{loc}(\mathfrak{A}_1 + \mathfrak{A}_2)$;

(vii) $V_1 \cup V_2 = \text{loc}(\mathfrak{A}_1 \cdot \mathfrak{A}_2) = \text{loc}(\mathfrak{A}_1 \cap \mathfrak{A}_2)$,

where $\mathfrak{A}_1 + \mathfrak{A}_2$ is the ideal consisting of sums of elements from the separate ideals, and $\mathfrak{A}_1 \cdot \mathfrak{A}_2$ is the ideal generated by products of elements from the two ideals.

A germ V of an analytic subvariety at the origin in $\mathbf{C}^n$ is said to be <u>reducible</u> if it can be written $V = V_1 \cup V_2$, where $V_i \subset V$ are also germs of analytic subvarieties at the origin in $\mathbf{C}^n$; a germ which is not reducible is said to be <u>irreducible.</u> It is easy to see that a germ V is irreducible if and only if $\operatorname{id} V$ is a prime ideal in ${}_n\mathcal{O}_0$. To a considerable extent the study of germs of analytic subvarieties can be reduced to the study of irreducible germs, in view of the following observation.

Theorem 1. Any germ of an analytic subvariety at the origin in $\mathbf{C}^n$ can be written uniquely as an irredundant union of finitely many irreducible germs of analytic subvarieties.

Proof. First suppose that there is a germ V of an analytic subvariety which cannot be written as a finite union of irreducible germs. Since V cannot itself then be irreducible, necessarily $V = V_1 \cup V_1'$ where V_1, V_1' are germs of analytic subvarieties and are properly contained in V ; and at least one of these two germs, say V_1 , in turn cannot be written as a finite union of irreducible germs. Repeating the argument, it follows that $V_1 = V_2 \cup V_2'$, where V_2, V_2' are germs of analytic subvarieties and are properly contained in V_1 , and V_2 cannot be written as a finite union of irreducible germs. Proceeding in this way, there results a strictly decreasing sequence of germs of analytic subvarieties $V \supset V_1 \supset V_2 \supset \dots$; and consequently there also results a strictly increasing sequence of ideals $\operatorname{id} V \subset \operatorname{id} V_1 \subset \operatorname{id} V_2 \subset \dots$ in the ring ${}_n\mathcal{O}_0$. This is impossible, since the ring ${}_n\mathcal{O}_0$ is Noetherian; and therefore every germ of an analytic subvariety can be written as a finite union of irreducible germs. Suppose next that a germ V of an analytic subvariety is written as a finite union of irreducible germs in two ways, say $V = V_1 \cup \dots \cup V_r =$ $= V_1' \cup \dots \cup V_s'$. It can of course be assumed that these are irredundant representations, in the sense that none of the germs V_i, V'_i can be omitted in these representations, or equivalently, that $V_i \not\subseteq \bigcup_{j\neq i} V_j$ and $V'_i \not\subseteq \bigcup_{j\neq i} V'_j$. Note that for any index i, $V_i = V_i \cap V = (V_i \cap V'_1) \cup \ldots \cup (V_i \cap V'_s)$; but since V_i is irreducible, necessarily $V_i = V_i \cap V'_{f(i)}$ and hence $V_i \subseteq V'_{f(i)}$ for some index $f(i)$. Similarly of course, for any index i it follows that $V'_i \subseteq V_{g(i)}$ for some index $g(i)$. Thus $V_i \subseteq V'_{f(i)} \subset V_{g(f(i))}$ and $V'_i \subseteq V_{g(i)} \subseteq V'_{f(g(i))}$; since the two representations are irredundant, it follows that $g(f(i)) = f(g(i)) = i$ and $V'_i = V_{g(i)}$. The two representations thus merely differ in the order in which the terms are written, and the proof is thereby concluded.

When a germ V of an analytic subvariety at the origin in $\mathbb{C}^n$ is written as an irredundant finite union of irreducible germs $V = V_1 \cup \ldots \cup V_r$, these germs V_i are called the irreducible components or irreducible branches of the germ V.

(b) Considering only germs of analytic subvarieties at the origin in $\mathbb{C}^n$ is of course merely a notational convenience; it is quite evident that a simple translation extends the preceding results to any other point of $\mathbb{C}^n$. Actually of course all of the preceding observations are clearly preserved under any complex analytic homeomorphism from an open neighborhood of the origin to an open neighborhood of any other point in $\mathbb{C}^n$, and in particular, under any complex analytic homeomorphism between two open neighborhoods of the origin in $\mathbb{C}^n$. The intrinsic properties of a germ of an analytic subvariety at the origin in $\mathbb{C}^n$ are independent of the choice of coordinates at the origin in $\mathbb{C}^n$.

However it is often useful to choose coordinates at the origin in $\mathbb{C}^n$ which are conveniently positioned for studying a particular germ of an analytic subvariety. A set of coordinates $z_1,\ldots,z_n$ at the origin in $\mathbb{C}^n$ is said to be a regular system of coordinates for an ideal $\mathfrak{A} \subset {}_n\mathcal{O}$ if for some integer $0 \leq k \leq n$:

(i) ${}_k\mathcal{O} \cap \mathfrak{A} = 0$;

(ii) ${}_{j-1}\mathcal{O}[z_j] \cap \mathfrak{A}$ contains a Weierstrass polynomial in z_j for $j = k+1,\ldots,n$.

The integer k is called the dimension of the ideal $\mathfrak{A}$ with respect to this system of coordinates. Note that the imbedding ${}_j\mathcal{O} \subseteq {}_n\mathcal{O}$ depends on the coordinate system, viewing an element $f \in {}_j\mathcal{O}$ as an element of ${}_n\mathcal{O}$ depending only on the first j coordinates.

Theorem 2. For any ideal $\mathfrak{A} \subset {}_n\mathcal{O}$ there is a regular system of coordinates at the origin in $\mathbb{C}^n$.

Proof. If $\mathfrak{A} = 0$ it is clear that any set of coordinates at the origin in $\mathbb{C}^n$ is a regular system of coordinates for $\mathfrak{A}$, with respect to which $\mathfrak{A}$ has dimension n. If $\mathfrak{A} \neq 0$, select any nontrivial element $f_n \in \mathfrak{A}$. After making a linear change of coordinates at the origin in $\mathbb{C}^n$ if necessary, the function f_n can be assumed to be regular in z_n ; then from the Weierstrass preparation theorem it follows that $f_n = u_n \cdot g_n$, where $u_n \in {}_n\mathcal{O}$

is a unit and $g_n \in {}_{n-1}\mathcal{O}[z_n]$ is a Weierstrass polynomial in z_n. Since u_n is a unit, $g_n = \frac{1}{u_n} \cdot f_n \in {}_{n-1}\mathcal{O}[z_n] \cap \mathfrak{M}$. If ${}_{n-1}\mathcal{O} \cap \mathfrak{M} = 0$, this set of coordinates is a regular system of coordinates for $\mathfrak{M}$, with respect to which $\mathfrak{M}$ has dimension $n-1$. If ${}_{n-1}\mathcal{O} \cap \mathfrak{M} \neq 0$, select any nontrivial element $f_{n-1} \in {}_{n-1}\mathcal{O} \cap \mathfrak{M}$. After a linear change of coordinates at the origin in $\mathbb{C}^n$ involving only the first $n-1$ coordinates (hence really a linear change of coordinates at the origin in $\mathbb{C}^{n-1}$), the function f_{n-1} can be assumed to be regular in z_{n-1}; then from the Weierstrass preparation theorem in ${}_{n-1}\mathcal{O}$ it follows that $f_{n-1} = u_{n-1} \cdot g_{n-1}$, where $u_{n-1} \in {}_{n-1}\mathcal{O}$ is a unit and $g_{n-1} \in {}_{n-2}\mathcal{O}[z_{n-1}]$ is a Weierstrass polynomial in z_{n-1}. Again $g_{n-1} \in {}_{n-2}\mathcal{O}[z_{n-1}] \cap \mathfrak{M}$; and $g_n \in {}_{n-1}\mathcal{O}[z_n] \cap \mathfrak{M}$ remains a Weierstrass polynomial in z_n. If ${}_{n-2}\mathcal{O} \cap \mathfrak{M} = 0$, this set of coordinates is a regular system of coordinates for $\mathfrak{M}$, with respect to which $\mathfrak{M}$ has dimension $n-2$. If ${}_{n-2}\mathcal{O} \cap \mathfrak{M} \neq 0$, the process continues in the obvious manner. Since $\mathfrak{M}$ is by assumption a proper ideal, it is evident that ${}_0\mathcal{O} \cap \mathfrak{M} = 0$; the process thus eventually stops, and the theorem is thereby proved.

It should be pointed out, as is clear from the proof of the preceding theorem, that given any ideal $\mathfrak{M} \subset {}_n\mathcal{O}$ and coordinate system $z_1, \ldots, z_n$ at the origin in $\mathbb{C}^n$, a suitable linear change of coordinates will transform the given coordinate system into a regular system of coordinates for the ideal $\mathfrak{M}$.

The condition that a set of coordinates $z_1,\dots,z_n$ be a regular system of coordinates for an ideal $\mathfrak{A} \subset {}_n\mathcal{O}$ has a simple algebraic interpretation in terms of the residue class ring ${}_n\mathcal{O}/\mathfrak{A}$, as follows. (One reference for the general algebraic results to be used in the sequel is van der Waerden, Modern Algebra, (Frederick Ungar, 1950), volumes I and II.) To set the notation, for any element $f \in {}_n\mathcal{O}$ let $\tilde{f}$ denote the image of this element in the residue class ring ${}_n\mathcal{O}/\mathfrak{A}$, and similarly, for any subring ${}_j\mathcal{O} \subseteq {}_n\mathcal{O}$ let ${}_j\tilde{\mathcal{O}}$ denote the image of this subring in the residue class ring ${}_n\mathcal{O}/\mathfrak{A}$; then in particular the image of a coordinate function z_j in the residue class ring ${}_n\tilde{\mathcal{O}} = {}_n\mathcal{O}/\mathfrak{A}$ will be denoted by $\tilde{z}_j$. The Weierstrass polynomial $g_n \in {}_{n-1}\mathcal{O}[z_n] \cap \mathfrak{A}$ can be written $g_n = z_n^r + a_1 z_n^{r-1} + \dots + a_r$ for some elements $a_i \in {}_{n-1}\mathcal{O}$; and since $g_n \in \mathfrak{A}$, it follows that $\tilde{g}_n = 0$. Consequently there results an equation in the residue class ring ${}_n\tilde{\mathcal{O}}$ of the form $\tilde{z}_n^r + \tilde{a}_1 \tilde{z}_n^{r-1} + \dots + \tilde{a}_r = 0$; the element $\tilde{z}_n \in {}_n\tilde{\mathcal{O}}$ is therefore integral over the subring ${}_{n-1}\tilde{\mathcal{O}} \subset {}_n\tilde{\mathcal{O}}$. Furthermore, it follows from the Weierstrass division theorem that any element $f \in {}_n\mathcal{O}$ can be written $f = g_n \cdot q + r$ for some polynomial $r \in {}_{n-1}\mathcal{O}[z_n]$; so in the residue class ring, $\tilde{f} = \tilde{r} \in {}_{n-1}\tilde{\mathcal{O}}[\tilde{z}_n]$. That is to say, ${}_n\tilde{\mathcal{O}} \cong {}_{n-1}\tilde{\mathcal{O}}[\tilde{z}_n]$ is an integral algebraic extension of the subring ${}_{n-1}\tilde{\mathcal{O}} \subset {}_n\tilde{\mathcal{O}}$, generated by the element $\tilde{z}_n$. Considering then the Weierstrass polynomial $g_{n-1} \in {}_{n-2}\mathcal{O}[z_{n-1}] \cap \mathfrak{A}$, it follows similarly that ${}_{n-1}\tilde{\mathcal{O}} \cong {}_{n-2}\tilde{\mathcal{O}}[\tilde{z}_{n-1}]$ is an integral algebraic extension of the subring ${}_{n-2}\tilde{\mathcal{O}} \subset {}_{n-1}\tilde{\mathcal{O}}$, generated by the element $\tilde{z}_{n-1}$; and so on down the line. Finally, since

${}_k\mathcal{O} \cap \mathfrak{M} = 0$, it is clear that ${}_k\tilde{\mathcal{O}} \cong {}_k\mathcal{O}$. Altogether, applying the theorem of transitivity of integral extensions, ${}_n\tilde{\mathcal{O}} = {}_k\tilde{\mathcal{O}}[\tilde{z}_{k+1},\ldots,\tilde{z}_n]$ <u>is an integral algebraic extension of the subring</u> ${}_k\tilde{\mathcal{O}} \cong {}_k\mathcal{O}$, <u>generated by the</u> n-k <u>elements</u> $\tilde{z}_{k+1},\ldots,\tilde{z}_n$. Conversely, if the residue class ring ${}_n\tilde{\mathcal{O}} = {}_n\mathcal{O}/\mathfrak{M}$ has this form, it follows readily that the coordinate system is regular for the ideal $\mathfrak{M}$ and that $\mathfrak{M}$ has dimension k with respect to this system of coordinates. For since ${}_k\tilde{\mathcal{O}} \cong {}_k\mathcal{O}$, necessarily ${}_k\mathcal{O} \cap \mathfrak{M} = 0$. Further, since $\tilde{z}_j$ is integral over ${}_k\tilde{\mathcal{O}} \cong {}_k\mathcal{O}$ for any value $j = k+1,\ldots,n$, there must exist a monic polynomial $p_j(X) \in {}_k\mathcal{O}[X]$ such that $\tilde{p}_j(\tilde{z}_j) = 0$ in ${}_n\tilde{\mathcal{O}}$, or equivalently such that $p_j(z_j) \in {}_k\mathcal{O}[z_j] \cap \mathfrak{M} \subseteq {}_{j-1}\mathcal{O}[z_j] \cap \mathfrak{M}$. Any such polynomial of the smallest degree must necessarily be a Weierstrass polynomial; for otherwise $p_j(z_j)$ would be regular in z_j of order less than its degree, and an application of the Weierstrass preparation theorem in the local ring $\mathbb{C}\{z_1,\ldots,z_k,z_j\}$ would yield a Weierstrass polynomial of still smaller degree in ${}_k\mathcal{O}[z_j] \cap \mathfrak{M}$.

If the coordinates $z_1,\ldots,z_n$ form a regular system of coordinates for an ideal $\mathfrak{M} \subset {}_n\mathcal{O}$, with respect to which the ideal $\mathfrak{M}$ has dimension k , it follows as in the preceding paragraph that there are Weierstrass polynomials $p_j \in {}_k\mathcal{O}[z_j] \cap \mathfrak{M}$ for $j = k+1,\ldots,n$. Choosing such polynomials of the smallest degree, the resulting set of n-k germs will be called a <u>first set of canonical equations for the ideal</u> $\mathfrak{M}$ <u>with respect to the given coordinate system.</u> Condition (ii) in the definition of a regular system of coordinates for an ideal can be replaced by the existence of a first set of canonical equations for the ideal with respect to the given coordinate system, often an easier condition to use.

Now to consider the geometrical significance of a set of coordinates $z_1,\ldots,z_n$ being a regular system of coordinates for an ideal $\mathfrak{A} \subset {}_n\mathcal{O}$, select a first set of canonical equations $p_{k+1},\ldots,p_n$ for the ideal with respect to the given coordinate system. Choose an open neighborhood U' of the origin in $\mathbb{C}^k$ such that the coefficients of these polynomials p_j are analytic throughout U'; the functions p_j can then be viewed as analytic functions in the open subset $U' \times \mathbb{C}^{n-k} \subseteq \mathbb{C}^n$, and these functions define an analytic subvariety $W = \{z \in U' \times \mathbb{C}^{n-k} | p_{k+1}(z) = \ldots = p_n(z) = 0\}$ in that open set. Since $p_j \in \mathfrak{A}$ it is evident that any analytic subvariety representing the germ $\operatorname{loc} \mathfrak{A}$ must be contained in the subvariety W in some open neighborhood of the origin, or equivalently, as germs $\operatorname{loc} \mathfrak{A} \subseteq W$. The subvariety W has a very simple description, and this makes it possible to say some things about the germ $\operatorname{loc} \mathfrak{A}$.

Theorem 3. If $z_1,\ldots,z_n$ form a regular system of coordinates for an ideal $\mathfrak{A} \subset {}_n\mathcal{O}$, with respect to which the ideal has dimension k, then there are arbitrarily small open product neighborhoods $U = U' \times U'' \subseteq \mathbb{C}^k \times \mathbb{C}^{n-k} = \mathbb{C}^n$ of the origin admitting analytic subvarieties $V \subseteq U$ representing the germ $\operatorname{loc} \mathfrak{A}$, such that the mapping $\pi\colon V \longrightarrow U'$ induced by the natural projection mapping $U' \times U'' \longrightarrow U'$ is a proper light, continuous mapping.

(Recall that the mapping π is said to be proper if $\pi^{-1}(K)$ is a compact subset of V whenever K is a compact subset of U', and light if $\pi^{-1}(z')$ is a discrete set of points for any point $z' \in U'$.)

Proof. Consider any open neighborhood U_0 of the origin in $\mathbf{C}^n$ admitting an analytic subvariety $V_0 \subseteq U_0$ representing the germ loc $\mathfrak{A}$; and assume that U_0 is sufficiently small that the first set of canonical equations $p_{k+1}, \ldots, p_n$ are analytic throughout U_0, and that $V_0 \subseteq W_0$ where
$W_0 = \{z \in U_0 | p_{k+1}(z) = \ldots = p_n(z) = 0\}$. If $U = U' \times U'' \subseteq U_0$ is any product subneighborhood, then since $p_j \in \mathcal{O}_{U'}[z_j]$ is a Weierstrass polynomial in z_j at the origin, the leading coefficient is identically 1 while the remaining coefficients are analytic functions of $z' \in U'$ which vanish at the origin; these other coefficients can then be made arbitrarily small by choosing U' small enough. In particular, choose U' sufficiently small that
$W \subseteq U' \times U''$, where
$W = \{z = (z', z_{k+1}, \ldots, z_n) | z' \in U',\ p_j(z', z_j) = 0 \text{ for } j = k+1, \ldots, n\}$;
and note that $V = V_0 \cap U \subseteq W_0 \cap U = W$, where V is a representative of the germ loc $\mathfrak{A}$ in the neighborhood U. It is clear that $(K \times \mathbf{C}^{n-k}) \cap W$ is a compact subset of $U' \times U''$ whenever K is a compact subset of U'; and since V is a relatively closed subset of W and $(K \times \mathbf{C}^{n-k}) \cap V = \pi^{-1}(K)$, it follows that the mapping π is proper. It is also clear that the mapping π is light and continuous, hence the theorem is proved.

It should be noted that the preceding proof really only used the existence of the Weierstrass polynomials $p_j \in {}_k\mathcal{O}[z_j] \cap \mathfrak{A}$ for $j = k+1,\ldots,n$; but since $p_j \in {}_k\mathcal{O}[z_j] \cap \mathfrak{A} \subseteq {}_\ell\mathcal{O}[z_j] \cap \mathfrak{A}$ for $j = \ell+1,\ldots,n$ whenever $k \leq \ell \leq n$, the same conclusion holds with k replaced by ℓ . Therefore the following is an immediate consequence of the same reasoning.

Corollary to Theorem 3. Under the hypotheses of Theorem 3, for any integer ℓ , $k \leq \ell \leq n$, there are arbitrarily small open product neighborhoods $U = U' \times U'' \subseteq \mathbf{C}^\ell \times \mathbf{C}^{n-\ell} = \mathbf{C}^n$ of the origin admitting analytic subvarieties $V \subseteq U$ representing the germ loc $\mathfrak{A}$, such that the mapping $\pi\colon V \longrightarrow U'$ induced by the natural projection mapping $U' \times U'' \longrightarrow U'$ is a proper, light, continuous mapping.

(c) Turning next to the special case that the ideal $\mathfrak{A} \subset {}_n\mathcal{O}$ is a prime ideal, a good deal more can be said both algebraically and geometrically. In order to keep firmly in mind the restriction that the ideal be prime, it will be denoted by $\mathfrak{P}$ throughout this discussion.

Beginning with the algebraic considerations, for a prime ideal $\mathfrak{P} \subset {}_n\mathcal{O}$ the residue class ring ${}_n\tilde{\mathcal{O}} = {}_n\mathcal{O}/\mathfrak{P}$ is an integral domain, hence has a quotient field ${}_n\tilde{\mathcal{M}}$. Since, when the coordinates $z_1,\ldots,z_n$ are a regular system of coordinates for the ideal $\mathfrak{P}$ with respect to which that ideal has dimension k , it was shown that ${}_n\tilde{\mathcal{O}} \cong {}_k\tilde{\mathcal{O}}[\tilde{z}_{k+1},\ldots,\tilde{z}_n]$, where the elements $\tilde{z}_{k+1},\ldots,\tilde{z}_n$ are algebraic over the ring ${}_k\tilde{\mathcal{O}} = {}_k\mathcal{O}$, then necessarily the quotient field has the form ${}_n\tilde{\mathcal{M}} \cong {}_k\tilde{\mathcal{M}}[\tilde{z}_{k+1},\ldots,\tilde{z}_n]$, where ${}_k\tilde{\mathcal{M}} \cong {}_k\mathcal{M}$ is the quotient field of the integral domain ${}_k\tilde{\mathcal{O}} \cong {}_k\mathcal{O}$. Each element $\tilde{z}_j$ is algebraic over the field ${}_k\tilde{\mathcal{M}}$, hence is the root of an irreducible polynomial over the field ${}_k\tilde{\mathcal{M}}$; if the leading coefficient of this polynomial is taken to be 1 the polynomial is uniquely determined, and will be called the defining equation for the element $\tilde{z}_j$. Now since the elements $\tilde{z}_j$ are actually integral over the ring ${}_k\tilde{\mathcal{O}} \cong {}_k\mathcal{O}$, all the coefficients of the defining equation are elements of ${}_k\tilde{\mathcal{M}} \cong {}_k\mathcal{M}$ integral over ${}_k\tilde{\mathcal{O}} \cong {}_k\mathcal{O}$, hence are elements of ${}_k\tilde{\mathcal{O}} \cong {}_k\mathcal{O}$, since ${}_k\mathcal{O}$ is a unique factorization domain so is integrally closed in its quotient field. (See van der Waerden, §101.) The defining equation is thus a monic polynomial $p_j(X) \in {}_k\mathcal{O}[X]$ of minimal degree such that $\tilde{p}_j(\tilde{z}_j) = 0$, or equivalently, such that $p_j(z_j) \in {}_k\mathcal{O}[z_j] \cap \mathcal{Y}$. Consequently the defining equations of the elements $\tilde{z}_{k+1},\ldots,\tilde{z}_n$ are just the first set of canonical equations for the ideal $\mathcal{Y}$ with respect to the given coordinate system; and from this observation it is apparent that for a prime ideal $\mathcal{Y}$ <u>the first set of canonical equations is uniquely determined</u> by the choice of the coordinate system.

It follows from the theorem of the primitive element that there are complex constants $c_{k+1},\ldots,c_n$ such that the single element $c_{k+1}\tilde{z}_{k+1} + \ldots + c_n\tilde{z}_n$ generates the entire field extension ${}_n\tilde{\mathcal{M}} = {}_k\tilde{\mathcal{M}}[\tilde{z}_{k+1},\ldots,\tilde{z}_n]$ over ${}_k\tilde{\mathcal{M}} \cong {}_k\mathcal{M}$. (See van der Waerden, §40.) By making a suitable linear change of coordinates in the space

$\mathbb{C}^{n-k}$ of the variables $z_{k+1},\ldots,z_n$, it can clearly be assumed that $\tilde{z}_{k+1}$ itself generates the entire field extension. It is obvious from the algebraic interpretation that the new coordinates will still be a regular system of coordinates for the ideal $\mathfrak{p} \subset {}_n\mathcal{O}$; a regular system of coordinates for a prime ideal $\mathfrak{p} \subset {}_n\mathcal{O}$ with the additional property that ${}_n\tilde{\mathcal{M}} \cong {}_k\tilde{\mathcal{M}}[\tilde{z}_{k+1}]$ will be called a <u>strictly regular system of coordinates for that ideal.</u> The preceding remarks show that any prime ideal $\mathfrak{p} \subset {}_n\mathcal{O}$ has a strictly regular system of coordinates. For any such system of coordinates, the field extension ${}_n\tilde{\mathcal{M}} \cong {}_k\tilde{\mathcal{M}}[\tilde{z}_{k+1}]$ over ${}_k\tilde{\mathcal{M}}$ is fully described by the single member p_{k+1} of the first set of canonical equations for the ideal $\mathfrak{p}$; and the ring extension ${}_n\tilde{\mathcal{O}} = {}_k\tilde{\mathcal{O}}[\tilde{z}_{k+1},\ldots,\tilde{z}_n]$ is almost fully described by that single canonical equation as well. To see this, recall that the discriminant d of the polynomial $p_{k+1} \in {}_k\mathcal{O}[z_{k+1}]$ is an element of the ring ${}_k\mathcal{O}$, and that if r is the degree of the polynomial $p_{k+1} \in {}_k\mathcal{O}[z_{k+1}]$, any element of the ring ${}_n\tilde{\mathcal{O}} \cong {}_k\tilde{\mathcal{O}}[\tilde{z}_{k+1},\ldots,\tilde{z}_n]$ can be written uniquely as a linear combination of the elements $1/d, z_{k+1}/d,\ldots,z_{k+1}^{r-1}/d$ with coefficients from the ring ${}_k\tilde{\mathcal{O}}$. (See van der Waerden, §101. This is not really a full description of the ring ${}_n\tilde{\mathcal{O}} \cong {}_k\tilde{\mathcal{O}}[\tilde{z}_{k+1},\ldots,\tilde{z}_n]$, since not all linear combinations of these elements necessarily lie in the ring ${}_n\tilde{\mathcal{O}}$.) Equivalently, for any element $f \in {}_n\mathcal{O}$ there is a unique polynomial $q_f(X) \in {}_k\mathcal{O}[X]$ of degree strictly less than r such that $d\cdot f + q_f(z_{k+1}) \in \mathfrak{p}$. In particular, to each coordinate function

z_j for $k+2 \leqq j \leqq n$ there corresponds a unique polynomial $q_{z_j}(X) \in {}_k\mathcal{O}[X]$ of degree strictly less than r such that

$$q_j = d \cdot z_j - q_{z_j}(z_{k+1}) \in {}_k\mathcal{O}[z_{k+1}, z_j] \cap \mathfrak{p} \ ;$$

the $n-k-1$ polynomials $q_{k+2}, \ldots, q_n$ will be called the <u>second set of canonical equations for the ideal</u> $\mathfrak{p}$ <u>with respect to the given coordinate system.</u> These canonical equations are also uniquely determined by the choice of the coordinate system; unlike the first set of canonical equations, the second set are only determined for a strictly regular system of coordinates for a prime ideal $\mathfrak{p} \subset {}_n\mathcal{O}_0$. The two sets of canonical equations together generate an ideal $\mathcal{C} \subset {}_n\mathcal{O}$ called the <u>canonical ideal</u> for the ideal $\mathfrak{p}$ with respect to the given coordinate system; and the canonical equations $p_{k+1}, q_{k+2}, \ldots, q_r$ generate an ideal $\mathcal{C}_1 \subset {}_n\mathcal{O}$ called the <u>restricted canonical ideal</u> for the ideal $\mathfrak{p}$ with respect to the given coordinate system. It is obvious that $\mathcal{C}_1 \subseteq \mathcal{C} \subseteq \mathfrak{p}$; and although these may be strict containments, the following result shows that these various ideals cannot really differ by very much.

<u>Theorem 4.</u> If $z_1, \ldots, z_n$ form a strictly regular system of coordinates for a prime ideal $\mathfrak{p} \subset {}_n\mathcal{O}$, with respect to which the ideal has dimension k , and if $\mathcal{C}$ and $\mathcal{C}_1$ are the canonical and restricted canonical ideals for the ideal $\mathfrak{p}$ with respect to these coordinates, then for some integers a, b,

$$d^a \mathfrak{p} \subseteq \mathcal{C} \subseteq \mathfrak{p} \quad \text{and} \quad d^b \mathcal{C} \subseteq \mathcal{C}_1 \subseteq \mathcal{C} \ ,$$

where $d \in {}_k\mathcal{O}$ is the discriminant of the canonical equation $p_{k+1} \in {}_k\mathcal{O}[z_{k+1}] \cap \mathcal{Y}$.

Proof. Since the canonical equations $p_j \in {}_k\mathcal{O}[z_j] \cap \mathcal{Y}$ are Weierstrass polynomials of degrees r_j, repeated application of the Weierstrass division theorem shows that any element $f \in {}_n\mathcal{O}$ can be written as a polynomial in ${}_k\mathcal{O}[z_{k+1},\ldots,z_n]$, of degree strictly less than r_j in the variable z_j, modulo the ideal generated by the first set of canonical equations; for f can be written as a multiple of p_n plus a polynomial in ${}_{n-1}\mathcal{O}[z_n]$ of degree strictly less than r_n, each coefficient in this polynomial can then be written as a multiple of p_{n-1} plus a polynomial in ${}_{n-2}\mathcal{O}[z_{n-1}]$ of degree strictly less than r_{n-1}, and so on. Since the expression $d^i \cdot z_j^i$ is equal to a polynomial in ${}_k\mathcal{O}[z_{k+1}]$ modulo a multiple of the second canonical equation q_j, for any positive integer i and any index $k+2 \leq j \leq n$, it follows that for large enough a the element $d^a \cdot f$ can be written as a polynomial in ${}_k\mathcal{O}[z_{k+1}]$ modulo the canonical ideal $\mathcal{L}$; and after dividing by p_{k+1} again, this final polynomial can be taken to have degree strictly less than r_{k+1}. That is to say, if a is sufficiently large, then for any element $f \in {}_n\mathcal{O}$ there exists a polynomial $p_f \in {}_k\mathcal{O}[z_{k+1}]$ of degree strictly less than r_{k+1} such that $d^a \cdot f - p_f \in \mathcal{L}$. If $f \in \mathcal{Y}$ it follows that $p_f \in \mathcal{Y}$ as well, since $\mathcal{L} \subseteq \mathcal{Y}$; but then necessarily $p_f = 0$, since r_{k+1} is the degree of the polynomial of least degree in ${}_k\mathcal{O}[z_{k+1}] \cap \mathcal{Y}$, so that $d^a \cdot f \in \mathcal{L}$. This shows that $d^a \cdot \mathcal{Y} \subseteq \mathcal{L}$. Now if in the preceding argument the element f is from the beginning an element of the polynomial ring ${}_k\mathcal{O}[z_{k+1},\ldots,z_n]$, the initial application of the Weierstrass division theorem is not needed; that is to say, for any element $f \in {}_k\mathcal{O}[z_{k+1},\ldots,z_n]$ there exist an integer b and a polynomial $p_f \in {}_k\mathcal{O}[z_{k+1}]$ of degree strictly less than r_{k+1} such that $d^b\cdot f - p_f \in \mathcal{L}_1$. As before then, whenever $f \in {}_k\mathcal{O}[z_{k+1},\ldots,z_n] \cap \mathcal{Y}$, there is an integer b such that $d^b\cdot f \in \mathcal{L}_1$. Applying this in particular to the canonical equations $p_{k+2},\ldots,p_n$, it follows that $d^b\cdot\mathcal{L} \subseteq \mathcal{L}_1$, and the proof is thereby concluded.

As a matter of minor interest, it might be noted that in the preceding proof one can take $a = \sum_{i=k+2}^{n} (r_i-1)$ and $b = \max(r_{k+2},\ldots,r_n)$.

(d) The geometric significance of a set of coordinates $z_1,\ldots,z_n$ being a strictly regular system of coordinates for a prime ideal $\mathcal{Y} \subset {}_n\mathcal{O}$ follows now from a comparison of the germ $\operatorname{loc}\mathcal{Y}$ with the germs $\operatorname{loc}\mathcal{L}$ and $\operatorname{loc}\mathcal{L}_1$ defined by the canonical equations for the ideal with respect to the given coordinate system. Suppose that the ideal $\mathcal{Y}$ has dimension k with respect to this coordinate system. Let $p_{k+1},\ldots,p_n$ and $q_{k+2},\ldots,q_n$ be the first and second sets of canonical equations for the ideal and $d \in {}_k\mathcal{O}$ be the discriminant of the polynomial p_{k+1} ; and select germs $f_1,\ldots,f_r \in {}_n\mathcal{O}$ which generate the ideal. Let U be an open neighborhood of the origin sufficiently small that all of these germs are represented by analytic functions in U ; it is convenient to take this neighborhood in the form of a product domain $U = U' \times U'' \times U''' \subseteq \mathbb{C}^k \times \mathbb{C}^1 \times \mathbb{C}^{n-k-1} = \mathbb{C}^n$. The subset

$$V = \{z \in U \mid f_1(z) = \ldots = f_r(z) = 0\}$$

is then an analytic subvariety of the open set U which represents the germ $\operatorname{loc} \mathcal{V}$. Since the canonical equations are actually polynomials in the last $n-k$ coordinates, they are actually analytic functions in all of $U' \times \mathbb{C}^{n-k}$; the subsets

$$W = \{z \in U' \times \mathbb{C}^{n-k} \mid p_{k+1}(z) = \ldots = p_n(z) = q_{k+2}(z) = \ldots = q_n(z) = 0\}$$

and

$$W_1 = \{z \in U' \times \mathbb{C}^{n-k} \mid p_{k+1}(z) = q_{k+2}(z) = \ldots = q_n(z) = 0\}$$

are then analytic subvarieties of the open set $U' \times \mathbb{C}^{n-k}$ which represent the germs $\operatorname{loc} \mathcal{W}$ and $\operatorname{loc} \mathcal{W}_1$ respectively. Since the first canonical equations are all Weierstrass polynomials, it follows as in the proof of Theorem 3 that U' can be chosen sufficiently small that $W \subset U$; the conclusions of that theorem then hold for the subvariety W , so that the mapping from W to U' induced by the natural projection is a proper, light, continuous mapping. Finally the discriminant d only depends on the first k variables, so defines an analytic subvariety

$$D = \{z' \in U' \mid d(z') = 0\}$$

of the open subset $U' \subseteq \mathbb{C}^k$.

The relations between the ideal $\mathfrak{P}$ and its associated canonical ideals $\mathfrak{K}$ and $\mathfrak{K}_1$ given in Theorem 4 can be expressed in terms of the chosen generators for these various ideals as follows. There exist germs h'_{ij}, h''_{ij}, a'_{ij}, a''_{ij}, b_i, b_{ij} in ${}_n\mathcal{O}$ such that

$$p_i = \sum_{j=1}^{r} h'_{ij} f_j , \quad (i = k+1,\ldots,n)$$

$$q_i = \sum_{j=1}^{r} h''_{ij} f_j , \quad (i = k+2,\ldots,n)$$

$$d^a f_i = \sum_{j=k+1}^{n} a'_{ij} p_j + \sum_{j=k+2}^{n} a''_{ij} q_j , \quad (i = 1,\ldots,r)$$

$$d^b p_i = b_i p_{k+1} + \sum_{j=k+2}^{n} b_{ij} q_j , \quad (i = k+2,\ldots,n) .$$

Suppose that the neighborhood U is also chosen sufficiently small that these additional germs are represented by analytic functions in U , and the preceding relations hold throughout U . It is then immediately evident that

$$V \subseteq W \subseteq W_1 ;$$

and indeed that

$$V \cap ((U'-D) \times U'' \times U''') = (W \cap ((U'-D) \times U'' \times U''') = W_1 \cap ((U'-D) \times U'' \times U''') ,$$

or equivalently, that these three subvarieties coincide outside of the closed subset $D \times U'' \times U''' \subset U' \times U'' \times U''') = U$.

Theorem 5. Let $z_1,\ldots,z_n$ form a strictly regular system of coordinates for the prime ideal $\mathfrak{P} \subset {}_n\mathcal{O}$, with respect to which the ideal has dimension k ; and consider the first canonical equation

$p_{k+1} \in {}_k\mathcal{O}[z_{k+1}] \cap \mathcal{Y}$, a Weierstrass polynomial with discriminant $d \in {}_k\mathcal{O}$. There exist arbitrarily small connected open product neighborhoods $U = U' \times U'' \times U''' \subseteq \mathbb{C}^k \times \mathbb{C}^1 \times \mathbb{C}^{n-k-1} = \mathbb{C}^n$ of the origin, and analytic subvarieties $V \subset U$, representing the germ $\text{loc}\,\mathcal{Y}$, and $V_0 = \{(z',z'') \in U' \times U'' | p_{k+1}(z',z'') = 0\} \subset U' \times U''$, with the following properties:

(i) The natural projection mapping $U' \times U'' \times U''' \longrightarrow U' \times U''$ induces a proper, light, continuous mapping $\pi\colon V \longrightarrow V_0$ with image all of V_0; and the natural projection mapping $U' \times U'' \longrightarrow U'$ in turn induces a proper, light, continuous mapping $\pi_0\colon V_0 \longrightarrow U'$ with image all of U'.

(ii) Introducing the analytic subvarieties

$$D = \{z' \in U' | d(z') = 0\}, \quad B = V \cap (D \times U'' \times U'''), \quad B_0 = V_0 \cap (D \times U''),$$

the restriction $\pi | V\text{-}B\colon V\text{-}B \longrightarrow V_0\text{-}B_0$ is a homeomorphism, and the restriction $\pi_0 | V_0\text{-}B_0\colon V_0\text{-}B_0 \longrightarrow U'\text{-}D$ is a finite-sheeted covering projection.

Remarks. For the definitions and properties of covering spaces see for instance E. H. Spanier, Algebraic Topology (McGraw-Hill, 1966). The subvarieties $B \subset V$ and $B_0 \subset V_0$ along which the projections may fail to be covering mappings will be called the critical loci of the subvarieties V and V_0 respectively, with respect to the given coordinate system. The configuration described in this theorem can perhaps most easily be kept in mind by referring to the following diagram:

$$
\begin{array}{rcccccc}
 & V - B & \subset V & \subset U' \times U'' \times U''' & \subseteq & \mathbb{C}^k \times \mathbb{C}^1 \times \mathbb{C}^{n-k-1} = \mathbb{C}^n \\
\text{(homeomorphism)} & \big\downarrow \pi | V\text{-}B & \big\downarrow \pi & \big\downarrow \text{(projection)} & & \\
 & V_0 - B_0 & \subset V_0 & \subset U' \times U'' & \subseteq & \mathbb{C}^k \times \mathbb{C}^1 = \mathbb{C}^{k+1} \\
\text{(finite covering)} & \big\downarrow \pi_0 | V\text{-}B_0 & \big\downarrow \pi_0 & \big\downarrow \text{(projection)} & & \\
 & U' - D & \subset U' & \subseteq U' & \subseteq & \mathbb{C}^k
\end{array}
$$

Proof. There are arbitrarily small connected open product neighborhoods $U = U' \times U'' \times U'''$ in which the constructions described in the paragraphs preceding the statement of the theorem can be carried out; and the desired results then follow directly from the obvious properties of the subvarieties W and W_1. It first follows from Theorem 3 that the mappings π and π_0 are proper, light, and continuous; for this, it is only necessary that the neighborhood U' be chosen sufficiently small that $W \subset U$. Next, since the first canonical equation p_{k+1} is one of the equations describing the subvariety V, it is clear that $\pi(z) \in V_0$ for any point $z \in V$. Conversely, for any point $(z',z'') \in V_0 - B_0$, since $d(z') \neq 0$ it is clear from the form of the second canonical equations that the relations $q_{k+2}(z',z'',z_{k+2}) = \ldots = q_n(z',z'',z_n) = 0$ determine the coordinates $z''' = (z_{k+2},\ldots,z_n)$ of the unique point $z = (z',z'',z''') \in W_1 \cap ((U'-D) \times \mathbb{C}^1 \times \mathbb{C}^{n-k-1}) = V-B$ for which $\pi(z) = (z',z'')$; the mapping $\pi|V\text{-}B\colon V\text{-}B \longrightarrow V_0 - B_0$ is therefore a one-to-one mapping from V-B onto V_0-B_0, hence a homeomorphism. Since π is proper, the image $\pi(V)$ is evidently the full subvariety V_0. Finally note that the subvariety $V_0 \subset U' \times U''$ can be defined as $V_0 = \{(z',z'') \in U' \times \mathbf{C} | p_{k+1}(z',z'') = 0\}$; so for each fixed point $a' \in U'-D$, the polynomial equation $p_{k+1}(a',z'') = 0$ has r distinct roots $z''_{(1)},\ldots,z''_{(r)}$ all lying in U'' , where r is the degree of the Weierstrass polynomial p_{k+1} . Applying the Weierstrass preparation theorem (or equivalently the implicit function theorem) to the germs defined by the function p_{k+1} in the local rings ${}_{k+1}\mathcal{O}_{(a',z''_{(i)})}$, it follows that in some open neighborhood $U'_{a'} \subseteq U'$ of the point a' there are r analytic functions $g_1(z'),\ldots,g_r(z')$ such that $g_i(a') = z'_{(i)}$ and that $V_0 \cap (U'_{a'} \times U'') = \pi_0^{-1}(U'_{a'})$ is the union of the r disjoint sets

$$\{(z',z'') \in \mathbf{C}^{k+1} | z' \in U'_{a'},\ z'' = g_i(z')\} \quad \text{for} \quad i = 1,\ldots,r\ .$$

Each of these sets is clearly mapped homeomorphically onto $U'_{a'}$ under the natural projection from $U' \times U''$ to U' ; and consequently the restriction $\pi_0 | V_0 - B_0 \colon V_0 - B_0 \longrightarrow U'-D$ is an r-sheeted covering projection in the usual sense. Since it is again clear that the image of π_0 is all of U' , the proof is thereby concluded.

There are several remarks about the preceding theorem and its proof which perhaps should be made here. First, it is clear that B_0 is really the branch locus of the mapping $\pi_0\colon V_0 \longrightarrow U'$ in the customary sense. For since V_0 is defined by the single Weierstrass polynomial equation $p_{k+1}(z',z'') = 0$, the points z' in the discriminant locus D are precisely the points for which the polynomial in z'' has fewer than r distinct roots. Thus the following result is immediate.

Corollary 1 to Theorem 5. With the hypotheses and notation of Theorem 5, the set V_0 is the point set closure in $U' \times U''$ of the r-sheeted covering space $V_0 - B_0$ over $U' - D$; and as a point $z' \in U' - D$ approaches a point of the discriminant locus D , some of the points $\{\pi_0^{-1}(z')\}$ in the covering space lying over z' approach coincidence.

Actually, since $p_{k+1}(z',z'')$ is a Weierstrass polynomial, all of the points $\{\pi_0^{-1}(z')\}$ approach the origin in U as the point z' approaches the origin in U' . Now in addition to this, note that the composition $\pi_0 \pi$ clearly exhibits $V-B$ as an r-sheeted covering space of $U'-D$ homeomorphic to the covering space $V_0 - B_0$. However in this case the set B is not necessarily the branch locus of the mapping $\pi_0 \pi \colon V \longrightarrow U'$ in the topological sense; for although some of the points of V_0 lying over a point $z' \in U' - D$ may approach coincidence as z' approaches a point of the discriminant locus D , the points of V lying over z' need not approach coincidence, and V may remain an r-sheeted covering space over some of the points of D . The proof that V is the point set closure of $V-B$ in U is somewhat more involved, and will be taken up separately shortly. Again note that all of the points of $\{\pi^{-1}\pi_0^{-1}(z')\}$ approach the origin in $U'' \times U'''$ as the point z' approaches the origin in U' .

Second, it was noted in the proof that in some open subneighborhood $U'_{a'} \subseteq U' - D$ of any point $a' \in U' - D$ there are r complex analytic functions which parametrize the r sheets of the covering space $\pi_0^{-1}(U'_{a'})$ over $U'_{a'}$; these r sheets are the disjoint complex analytic subvarieties of $U'_{a'} \times U''$ defined by the equations $z'' - g_i(z') = 0$ for the various values $i = 1,\ldots,r$. Actually it is clear that these sheets are k-dimensional complex analytic submanifolds of $U'_{a'} \times U''$; for introducing new complex analytic coordinates in some open neighborhood of any point of $U'_{a'} \times U''$ defined by $w_1 = z_1,\ldots,w_k = z_k$, $w_{k+1} = z_{k+1} - g_i(z_1,\ldots,z_k)$, the subvariety is locally just the coordinate hyperplane $w_{k+1} = 0$. (Assuming that the reader is familiar with the notions of differentiable manifolds and submanifolds, it suffices to remark that complex analytic manifolds and submanifolds are the obvious analogues; the only point of possible difficulty which must be kept in mind is that a k-dimensional complex analytic manifold or submanifold is a 2k-dimensional topological manifold or submanifold.) Furthermore, since the second canonical equations exhibit the last n-k-1 coordinates of a point $z \in V-B$ as complex analytic functions of the first k+1 coordinates $(z',z'') \in V_0-B_0$, it is evident that the r sheets of the covering space $\pi^{-1}\pi_0^{-1}(U'_{a'})$ are likewise k-dimensional complex analytic submanifolds of $U'_{a'} \times U'' \times U'''$ parametrized by some complex analytic maps $G_i\colon U'_{a'} \longrightarrow U'_{a'} \times U'' \times U'''$. Thus there results the following assertion.

Corollary 2 to Theorem 5. With the hypotheses and notation of Theorem 5, V_0-B_0 is a k-dimensional complex analytic submanifold of $(U'-D) \times U''$ and $V-B$ is a k-dimensional complex analytic submanifold of $(U'-D) \times U'' \times U'''$.

Finally, there is no loss of generality in assuming that the open set U is actually a complete product domain $U = U_1 \times \ldots \times U_n$, where U_i is an open neighborhood of the origin in the plane of the complex variable z_i; thus $U' = U_1 \times \ldots \times U_k$, $U'' = U_{k+1}$, and $U''' = U_{k+2} \times \ldots \times U_n$. For any index $1 \leq \ell \leq n$ the natural projection $U_1 \times \ldots \times U_n \longrightarrow U_1 \times \ldots \times U_\ell$ induces a mapping $\pi_\ell \colon V \longrightarrow U_1 \times \ldots \times U_\ell$ by restriction; thus in the previous notation, $\pi = \pi_{k+1}$ and $\pi_0 \pi = \pi_k$. As was already noted after the proof of Theorem 3, for any index $k \leq \ell \leq n$ the mapping π_ℓ is a proper, light, continuous mapping from V into $U_1 \times \ldots \times U_\ell$. It is indeed clear that the following also holds, as an immediate consequence of the parametrization noted in deriving Corollary 2.

<u>Corollary 3 to Theorem 5.</u> If in addition to the hypotheses of Theorem 5 the domain U is a complete product domain, then for any index $k+1 \leq \ell \leq n$ the restriction $\pi_\ell | V-B$ is a homeomorphism from $V-B$ onto its image $\pi_\ell(V-B)$ in $U_1 \times \ldots \times U_\ell$; and this image $\pi_\ell(V-B)$ is a complex analytic submanifold of $(U'-D) \times U_{k+1} \times \ldots \times U_\ell$ which is an r-sheeted covering space over $U'-D$ under the natural projection $(U'-D) \times U_{k+1} \times \ldots \times U_\ell \longrightarrow U'-D$.

There still remains the critical locus B to be considered in more detail. The canonical equations do not suffice to describe this subvariety fully, since B can be a proper subset of $W \cap (D \times U'' \times U''')$. However for many purposes a sufficiently complete description of the critical locus B is given by the following continuation of the preceding results.

Theorem 5. (continued) Suppose that V is an irreducible germ of a proper analytic subvariety at the origin in $\mathbb{C}^n$, and consider the prime ideal $\mathcal{Y} = \text{id}\, V \subset {}_n\mathcal{O}$. Then with the same notation as in the first part of the theorem it further follows that:

(iii) The subvariety V is the point set closure of $V-B$ in U.

Proof. Denoting the point set closure of $V-B$ in U by $\overline{V-B}$, it is clear that $\overline{V-B} \subseteq V$. To prove the theorem it suffices merely to show that $\overline{V-B}$ is itself an analytic subvariety of U; for then since $V = B \cup (\overline{V-B})$ and V is an irreducible germ at the origin, necessarily $V = \overline{V-B}$ in some neighborhood of the origin. Now to each analytic function $f \in \mathcal{O}_U$ associate a polynomial $p_f = p_f(z';X) \in \mathcal{O}_{U'}[X]$ in the following manner. For any point $z' \in U'-D$ there are r distinct points of V lying over z' under the covering projection $\pi_0\pi\colon V-B \longrightarrow U'-D$; label these points $G_1(z'),\ldots,G_r(z')$ in some order, recalling from the discussion of Corollary 2 that the mappings $G_i(z')$ can be chosen to be complex analytic functions of z' in some open neighborhood of any point of $U'-D$. The expression

$$p_f(z';X) = \prod_{i=1}^{n} (X - f(G_i(z')))$$

is a polynomial of degree r in the variable X, with the elementary symmetric functions of the values $f(G_i(z'))$, $i = 1,\ldots,r$, as coefficients; so it is evident that these coefficients are well defined complex analytic functions in all of $U'-D$. For any compact subset $K \subset U'$, the inverse image $\pi^{-1}\pi_0^{-1}(K)$ is a compact subset of U since both mappings π and π_0 are proper; consequently the values $f(G_i(z'))$ are uniformly bounded in $K \cap (U'-D)$, as are the elementary symmetric functions of these values. It follows from the generalized Riemann removable singularities theorem that the coefficients of the polynomial p_f extend to analytic functions in all of U', so that $p_f \in \mathcal{O}_{U'}[X]$. This polynomial has the properties that its degree is the number of sheets in the covering projection $\pi_0\pi\colon V-B \longrightarrow U'-D$, and that for each point $z' \in U'-D$ the roots of the polynomial equation $p_f(z';X) = 0$ are precisely the r values $X = f(G_i(z'))$ for $i = 1,\dots,r$. The composite function $p_f(z';f(z))$ is then an analytic function in $\mathcal{O}_U$ which vanishes on $V-B$ and hence on $\overline{V-B}$. Introduce the subset

$$V^* = \{z \in U \mid p_f(z';f(z)) = 0 \quad \text{for all} \quad f \in \mathcal{O}_U\} .$$

This is an analytic subvariety of U; for it follows from the corollary to the extended Weierstrass division theorem noted in §1(a) that a finite number of the functions $p_f(z';f(z))$ serve to define the subset V^* in some open neighborhood of any point of U. It is clear that $V^* \supseteq \overline{V-B}$; and the proof will be concluded by showing that $V^* \subseteq \overline{V-B}$. Consider any point $a = (a',a'',a''') \in U$ for which $a \notin \overline{V-B}$. There are only finitely many points $b_1,\dots,b_s \in \overline{V-B} \subseteq V$ lying over a' under the light proper mapping $\pi_0\pi\colon V \longrightarrow U'$, and all are distinct from the point a. Choose an analytic function $f \in \mathcal{O}_U$ such that $f(a) \neq f(b_k)$ for $k = 1,\dots,s$, and consider further the polynomial $p_f(z';X) \in \mathcal{O}_{U'}[X]$. If $a' \notin D$ then necessarily $s = r$ and the points b_k are just the points $G_i(a')$ in some order. The roots of the polynomial equation $p_f(a';X) = 0$ are precisely the r values $X = f(b_k)$; hence $p_f(a';f(a)) \neq 0$, so that $a \notin V^*$. If $a' \in D$ then select a sequence of points $a_j' \in U'-D$ converging to a' . The roots of the polynomial equation $p_f(a_j';X) = 0$ are just the r values $f(G_i(a_j'))$, and these approach the roots of the polynomial equation $p_f(a';X) = 0$ as a_j' approaches a' ; so for any root X_ℓ of the polynomial equation $p_f(a';X) = 0$ there will be some sequence of values $f(G_{i_j}(a_j'))$ approaching X_ℓ . Since the mapping $\pi_0\pi$ is proper, after passing to a subsequence if necessary the points $G_{i_j}(a_j')$ will converge to some limiting value, which must necessarily be one of the points b_k ; and hence $X_\ell = b_k$ for that value of the index k . Again $p_f(a';f(a)) \neq 0$, so that $a \notin V^*$. This suffices to verify that $V^* \subseteq \overline{V-B}$, and the proof is thereby concluded.

Again there are some remarks about the proof of this final part of the theorem which should be made here. Note that the essential element of the proof was the observation that $\overline{V-B}$ is an analytic subvariety of U ; and that the proof of this assertion really used only the conditions that $\overline{V-B}$ is an analytic submanifold of $(U'-D) \times U'' \times U'''$, and that the mapping $\overline{V-B} \longrightarrow U'$ induced by the natural projection $U' \times U'' \times U''' \longrightarrow U'$ is a light, proper mapping exhibiting $V-B$ as a covering space of $U'-D$. (This result is typical of a class of extension theorems for complex analytic subvarieties, theorems providing sufficient conditions for an analytic subvariety $V-B \subset (U' \times U'' \times U''' - D \times U'' \times U''')$ to extend by closure to an analytic subvariety of $U' \times U'' \times U'''$. This aspect is best left to a later, more general discussion.) Now on the one hand, this argument can also be applied to each connected component of the set $V-B$. That is to say, if $V-B = W_1 \cup \ldots \cup W_s$ where W_k are the connected components, then each set W_k satisfies the conditions under which this argument goes through; there are of course only finitely many connected components, since each must be a covering space of the connected open set $U'-D$. The closure $\overline{W}_k$ of the component W_k in U is an analytic subvariety of U; and it is evident that this subvariety contains the origin. The germ of the subvariety V at the origin is hence the union of the germs of these subvarieties $\overline{W}_k$; but since V is irreducible, necessarily $s = 1$. The following is therefore an immediate consequence.

<u>Corollary 4 to Theorem 5.</u> With the hypotheses and notation of Theorem 5 and its continuation, $V-B$ is a connected point set.

On the other hand, when U is a complete product domain $U = U_1 \times \ldots \times U_n$ this argument can also be applied to each projection $\pi_\ell(V-B)$ of the set $V-B$ into the factor $U_1 \times \ldots \times U_\ell$, for $\ell = k+1,\ldots,n$. The point set closure $\pi_\ell(\overline{V-B})$ is therefore an analytic subvariety of $U_1 \times \ldots \times U_\ell$ for $\ell = k+1,\ldots,n$. Since π_ℓ is continuous, $\overline{\pi_\ell(V-B)} \supseteq \pi_\ell(V)$; and since π_ℓ is proper, $\pi_\ell(V)$ is closed and hence $\overline{\pi_\ell(V-B)} \subseteq \pi_\ell(V)$. The following result is therefore a further consequence.

Corollary 5 to Theorem 5. If in addition to the hypotheses of Theorem 5 and its continuation the domain U is a complete product domain, then for any index $k+1 \leq \ell \leq n$ the image $\pi_\ell(V)$ is an analytic subvariety of $U_1 \times \dots \times U_\ell$.

Note that after an arbitrary nonsingular linear change of coordinates in $\mathbb{C}^n$ involving only the variables $z_{k+1}, \dots, z_n$, the total projection $\pi_k \colon V \longrightarrow \pi_k(V) = U' \subseteq \mathbb{C}^k$ is unchanged; hence the restriction of this projection to the inverse image of the subset $U'-D$ remains an r-sheeted unbranched covering of $U'-D$. In some open neighborhood $U'_{a'}$ of each point $a' \in U'-D$ there will be r analytic mappings $G_i \colon U'_{a'} \longrightarrow \mathbb{C}^{n-k}$ which parametrize the r sheets of this covering; and the partial projections $\pi_\ell(V) \subset \mathbb{C}^\ell$ for $k+1 \leq \ell \leq n$ are parametrized by the appropriate sets of components of these mappings. When not all the components of the mappings G_i are considered, it is possible that the images of different mappings G_i either coincide completely or intersect nontrivially. After choosing a larger analytic subvariety $D \subseteq D^* \subset U'$ if necessary, it can be assumed that the images of different mappings G_i are either disjoint or coincident in $U'-D^*$, even when considering only some of the components. Thus the partial projections $\pi_\ell(V) \subset \mathbb{C}^\ell$ are also unbranched coverings of $U'-D^*$, although possibly coverings with fewer than r sheets. These projections are then complex analytic submanifolds of $(U'-D^*) \times \mathbb{C}^{\ell-k}$, and as before, their closures are complex analytic subvarieties in a neighborhood of the origin in the image space. Consequently, even for coordinates which form a regular but not necessarily strictly regular system of coordinates for the ideal id $V \subset {}_n\mathcal{O}$, the partial projections $\pi_\ell(V)$ are analytic subvarieties of open subsets in $\mathbb{C}^\ell$. This can be summarized as follows.

<u>Corollary 6 to Theorem 5.</u> If V is an irreducible germ of a proper analytic subvariety at the origin in $\mathbb{C}^n$, and if $z_1,\ldots,z_n$ form a regular (but not necessarily strictly regular) system of coordinates for the prime ideal $\mathfrak{P} = \text{id } V$ with respect to which the ideal has dimension k, then for any index $k+1 \leq \ell \leq n$ the partial projection $\pi_\ell(V)$ of V is an irreducible germ of a proper analytic subvariety at the origin in $\mathbb{C}^\ell$.

Finally recall that the monic polynomial $p_f(z';X) \in \mathcal{O}_{U'}[X]$ constructed during the proof of the last part of the theorem has the property that $p_f(z';f(z)) = 0$ for any point $z = (z',z'',z''') \in \overline{V-B} = V$. Considering the germs of these various functions at the origin and passing to the residue class ring ${}_n\tilde{\mathcal{O}} = {}_n\mathcal{O}/\mathfrak{P}$ modulo the prime ideal $\mathfrak{P} = \text{id } V$, this polynomial determines an element $\tilde{p}_f(X) \in {}_k\tilde{\mathcal{O}}[X] \cong {}_k\mathcal{O}[X]$; and further, $\tilde{p}_f(\tilde{f}) = 0$ in ${}_n\tilde{\mathcal{O}} = {}_n\mathcal{O}/\mathfrak{P}$. Thus the polynomial $\tilde{p}_f(X)$ is the polynomial exhibiting $\tilde{f} \in {}_n\tilde{\mathcal{O}} = {}_n\mathcal{O}/\mathfrak{P}$ as an integral algebraic element over ${}_k\tilde{\mathcal{O}} = {}_k\mathcal{O} \subseteq {}_n\tilde{\mathcal{O}}$. This observation may help to clarify the geometrical significance of the earlier algebraic constructions, or the algebraical significance of these later geometric constructions.

Theorem 5 in its entirety, together with its various Corollaries, provides a very useful local picture of the complex analytic subvariety defined by a prime ideal $\mathcal{P} \subset {}_n\mathcal{O}$; this picture will be referred to as the <u>local parametrization theorem</u> for germs of analytic subvarieties. The next step is to derive some properties of germs of analytic subvarieties following readily from the local parametrization theorem.

§3. Some applications of the local parametrization theorem.

(a) One rather direct application of the local parametrization theorem is to the completion of the list given in §2(a) of elementary relations between germs of analytic subvarieties at the origin in $\mathbf{C}^n$ and ideals in the local ring ${}_n\mathcal{O}$. It was noted there that for any ideal $\mathfrak{A} \subseteq {}_n\mathcal{O}$ there is a containment relation $\mathfrak{A} \subseteq \text{id loc}\, \mathfrak{A}$; the question when this is really an equality was left open.

Theorem 6. For any prime ideal $\mathfrak{P} \subset {}_n\mathcal{O}$ it follows that $\mathfrak{P} = \text{id loc}\, \mathfrak{P}$.

Proof. It is of course only necessary to show that $\text{id loc}\, \mathfrak{P} \subseteq \mathfrak{P}$. Choose a strictly regular system of coordinates $z_1, \ldots, z_n$ for the ideal $\mathfrak{P}$ with respect to which the ideal has dimension k; and introduce the canonical equations for the ideal, and the other notation as in §2. For any element $f \in {}_n\mathcal{O}$, use of the canonical equations and repeated application of the Weierstrass division theorem as in the proof of Theorem 4 show that there is a polynomial $p_f \in {}_k\mathcal{O}[z_{k+1}]$ of degree strictly less than the degree r_{k+1} of the canonical equation $p_{k+1} \in {}_k\mathcal{O}[z_{k+1}]$, such that $d^a f - p_f \in \mathfrak{P}$ for some integer a. If $f \in \text{id loc}\, \mathfrak{P}$, then the polynomial p_f also vanishes on the analytic subvariety $V = \text{loc}\, \mathfrak{P}$ in $\mathbf{C}^n$; indeed, since $p_f \in {}_{k+1}\mathcal{O}$, this polynomial actually vanishes on the projection V_o of the subvariety V in the space $\mathbf{C}^{k+1}$, as described in the local parametrization theorem. Now for each point $z' \in U' - D$ there are r_{k+1} points $(z', z_{k+1}) \in V_o - B_o$

lying over z' under the covering projection $V_o - B_o \longrightarrow U' - D$; and since the polynomial $p_f(z', z_{k+1})$ has degree strictly less than (r_{k+1}) but vanishes at all these points, it is necessarily the zero polynomial. Thus $p_f = 0$, so that $d^a f \in \mathcal{Y}$; but since $\mathcal{Y}$ is prime and $d \notin \mathcal{Y}$, it follows that $f \in \mathcal{Y}$. This therefore shows that id loc $\mathcal{Y} \subseteq \mathcal{Y}$, and concludes the proof.

It should be noted that the proof of this result only required the use of the first part of Theorem 5. The proof of the final part of Theorem 5 required the additional hypothesis that the prime ideal $\mathcal{Y}$ be the ideal of an irreducible germ of a subvariety at the origin in $\mathbf{C}^n$; however, in view of Theorem 6, any prime ideal is such an ideal, and this additional hypothesis is therefore automatically satisfied. That is, <u>all of Theorem</u> 5 <u>holds for an arbitrary prime ideal</u> $\mathcal{Y} \subset {}_n\mathcal{O}$.

Now the treatment of the analogue of Theorem 6 for an arbitrary ideal $\mathfrak{M} \subset {}_n\mathcal{O}$ follows quite easily from the preceding result for the special case of a prime ideal, upon using a simple bit of additional algebraic machinery. Recall from the Lasker-Noether decomposition theorem that any ideal $\mathfrak{M}$ in the Noetherian ring ${}_n\mathcal{O}$ can be written as the intersection of a finite number of primary ideals. (See van der Waerden, §87.) The radical of an ideal $\mathfrak{M} \subseteq {}_n\mathcal{O}$ is the set

$$\sqrt{\mathfrak{M}} = \{f \in {}_n\mathcal{O} \mid f^r \in \mathfrak{M} \text{ for some integer } r > 0 \text{ depending on } f\};$$

clearly $\sqrt{\mathfrak{M}}$ is also an ideal in ${}_n\mathcal{O}$, and $\mathfrak{M} \subseteq \sqrt{\mathfrak{M}}$. The radical of a primary ideal is a prime ideal; and when an ideal $\mathfrak{M}$ is written as an intersection $\mathfrak{A} = \mathfrak{Q}_1 \cap \ldots \cap \mathfrak{Q}_r$ of primary ideals, its radical is the intersection $\sqrt{\mathfrak{A}} = \sqrt{\mathfrak{Q}_1} \cap \ldots \cap \sqrt{\mathfrak{Q}_r}$ of prime ideals.

Theorem 6 (continued). For any ideal $\mathfrak{A} \subset {}_n\mathcal{O}$ it follows that $\sqrt{\mathfrak{A}} = \text{id loc } \mathfrak{A}$.

Proof. If $\mathfrak{Q} \subset {}_n\mathcal{O}$ is a primary ideal, its radical $\sqrt{\mathfrak{Q}}$ is a prime ideal, and it is evident that $\text{loc} \sqrt{\mathfrak{Q}} = \text{loc } \mathfrak{Q}$; it then follows from the first part of Theorem 6 that $\text{id loc } \mathfrak{Q} = \text{id loc} \sqrt{\mathfrak{Q}} = \sqrt{\mathfrak{Q}}$. For any ideal $\mathfrak{A} \subset {}_n\mathcal{O}$ written as an intersection of primary ideals $\mathfrak{A} = \mathfrak{Q}_1 \cap \ldots \cap \mathfrak{Q}_r$, note that $\text{loc } \mathfrak{A} = \text{loc } \mathfrak{Q}_1 \cup \ldots \cup \text{loc } \mathfrak{Q}_r$ and $\text{id loc } \mathfrak{A} = \text{id loc } \mathfrak{Q}_1 \cap \ldots \cap \text{id loc } \mathfrak{Q}_r$; hence, as just observed, $\text{id loc } \mathfrak{A} = \sqrt{\mathfrak{Q}_1} \cap \ldots \cap \sqrt{\mathfrak{Q}_r} = \sqrt{\mathfrak{A}}$, and the proof is thereby concluded.

This result is usually called the <u>Hilbert zero theorem</u> (<u>Hilbertsche Nullstellensatz</u>). Another way of stating it is that if $f \in {}_n\mathcal{O}$ vanishes on $\text{loc } \mathfrak{A}$ for any ideal $\mathfrak{A} \subset {}_n\mathcal{O}$, then $f^r \in \mathfrak{A}$ for some integer $r > 0$. As a corollary, the ideals $\mathfrak{A} \subset {}_n\mathcal{O}$ for which $\mathfrak{A} = \text{id loc } \mathfrak{A}$ can be characterized purely algebraically as the radical ideals, those ideals $\mathfrak{A}$ such that $\mathfrak{A} = \sqrt{\mathfrak{A}}$.

(b) A slightly subtler application of the local parametrization theorem leads to a proof of the coherence of the sheaf of ideals of an analytic subvariety. This result can be stated quite simply without using sheaves, as follows; but the proof seems to require Oka's theorem or something of comparable depth, so sheaves will appear in the proof in order to effect some simplification.

Theorem 7. Suppose that V is an analytic subvariety of an open neighborhood U of the origin in $\mathbb{C}^n$, such that the germ of V at the origin is irreducible; and that $f_1,\ldots,f_r \in {}_n\mathcal{O}_U$ are analytic functions in U, such that their germs at the origin generate $\mathrm{id}\, V \subset {}_n\mathcal{O}_o$. Then if U is sufficiently small, the germs of the functions $f_1,\ldots,f_r \in {}_n\mathcal{O}_a$ at any point $a \in U$ generate the ideal $\mathrm{id}\, V \subseteq {}_n\mathcal{O}_a$ of the germ of the subvariety V at that point.

Proof. Again choose a strictly regular system of coordinates $z_1,\ldots,z_n$ for the ideal $\mathrm{id}\, V \subset {}_n\mathcal{O}_o$ with respect to which that ideal has dimension k; and introduce the canonical equations for the ideal, and the other notation as in §2. Assume that the neighborhood U is chosen sufficiently small that the canonical equations are analytic throughout U and are in the ideal in ${}_n\mathcal{O}_U$ generated by the functions $f_1,\ldots,f_r$; and that the local parametrization theorem holds in U. For any point $a \in U$ the local ring ${}_n\mathcal{O}_a$ is of course isomorphic to the local ring ${}_n\mathcal{O}_o$ and this isomorphism can be effected by the change of variable $w_i = z_i - a_i$; elements of ${}_n\mathcal{O}_a \cong {}_n\mathcal{O}_o$ will be written either as $g(z)$ or as $g(w)$ to indicate in which local ring they are to be considered as lying. It is clear that for $j = k+1,\ldots,n$ the germ of the canonical equation p_j at the point a is regular in the local coordinate w_j, since that germ is a nontrivial monic polynomial in ${}_k\mathcal{O}_o[w_j]$; hence applying the Weierstrass preparation theorem write $p_j(w) = p_j'(w)\cdot p_j''(w)$ where $p_j'(w) \in {}_k\mathcal{O}_o[w_j]$ is a Weierstrass polynomial in w_j and $p_j''(w) \in {}_n\mathcal{O}_o$ is a unit. It is also clear that for $j = k+2,\ldots,n$

the germ of the canonical equation q_j at the point a can be written $q_j(w) = d(w)\cdot w_j - q_j'(w)$ where $q_j'(w) \in {}_k\mathcal{O}_o[w_{k+1}]$ and $d(w) \in {}_k\mathcal{O}_o$ is as usual the germ of the discriminant of the polynomial $p_{k+1}(z)$. Note that since $p_j''(w)$ are units, the elements $p_{k+1}'(w),\ldots,p_n'(w),\ q_{k+2}'(w),\ldots,q_n'(w)$ all lie in the ideal in ${}_n\mathcal{O}_a \cong {}_n\mathcal{O}_o$ generated by the germs of the functions $f_1,\ldots,f_r$. For any element $g(w) \in {}_n\mathcal{O}_o$, use of the polynomials $p_{k+1}'(w),\ldots,p_n'(w),\ q_{k+2}'(w),\ldots,q_n'(w)$ and repeated application of the Weierstrass division theorem as in the proof of Theorem 4 show that there is a polynomial $p_g(w) \in {}_k\mathcal{O}_o[w_{k+1}]$ of degree strictly less than the degree of the polynomial $p_{k+1}''(w)$, such that for some integer s, $d(w)^s\cdot g(w) - p_g(w)$ is contained in the ideal in ${}_n\mathcal{O}_a \cong {}_n\mathcal{O}_o$ generated by the germs of the functions $f_1,\ldots,f_r$.

If $g(w)$ vanishes on the germ of the subvariety V at the point a, so does the polynomial $p_g(w)$; but one cannot conclude from this, as one did in the corresponding case in the proof of Theorem 6, that the polynomial $p_g(w)$ vanishes identically. The problem is that under the projection $\pi\colon V \longrightarrow V_o$ from $V \subset \mathbb{C}^n$ onto $V_o \subset \mathbb{C}^{k+1}$, the point $a \in V$ need not be the only point of V having the same image $\pi(a) = (a', a'') \in V_o \subset \mathbb{C}^k \times \mathbb{C}$ when a is contained in the critical locus $B \subset V$; the equation $p_{k+1}'(w) = 0$ defines the entire subvariety V_o in a neighborhood of the point $(a', a'') \in V_o$, while the polynomial $p_g(w)$ need only vanish on that part of V_o which is the image of a neighborhood of the point $a \in V$ under the mapping π. This can of course only happen when the covering space

$V_o - B_o$ over $U' - D$ is not connected near the point $(a', a'') \in V_o$; restricting the covering projections $\pi: V-B \longrightarrow V_o - B_o$ and $\pi_o: V_o - B_o \longrightarrow U'-D$ to the inverse images of a small open neighborhood of $a' \in U'$, some local components of $V_o - B_o$ will arise from local components of $V-B$ near the point $a \in V$, while others will arise from local components of $V-B$ near the other points of V mapping under π to the point $(a', a'') \in V_o$. As noted previously in corresponding situations, the closure of each connected component of the local covering $V_o - B_o$ near the point (a', a'') will be an analytic subvariety of an open neighborhood of that point; and it follows readily that the polynomial $p'_{k+1}(w)$ can be factored into a product $p'_{k+1}(w) = p^*_{k+1}(w) \cdot p^{**}_{k+1}(w)$, where $p^*_{k+1}(w) \in {}_k\mathcal{O}_o[w_{k+1}]$ is the defining equation for that part of V_o near the point (a', a'') arising as the image of a neighborhood of a in V . The remaining term $p^{**}_{k+1}(w)$ is of course a unit in the local ring ${}_n\mathcal{O}_a$, so that $p^*_{k+1}(w)$ is contained in the ideal ${}_n\mathcal{O}_a$ generated by the germs of the functions $f_1, \ldots, f_r$. Yet another application of the Weierstrass division theorem shows that $p_g(w) = p^*_{k+1}(w) \cdot q_g(w) + p^*_g(w)$ where the polynomial $p^*_g(w) \in {}_k\mathcal{O}_o[w_{k+1}]$ has degree strictly less than the degree of the polynomial $p^*_{k+1}(w)$. Now if $g(w)$ and hence $p_g(w)$ vanish on the germ of the subvariety V at the point a , it does follow as usual that $p^*_g(w) = 0$; and therefore the product $d(w)^s \cdot g(w)$ is contained in the ideal in ${}_n\mathcal{O}_a \cong {}_n\mathcal{O}_o$ generated by the germs of the functions $f_1, \ldots, f_r$ at the point a . (Note that the exponent s can really be bounded independently of the choice of the point a .)

At each point $a \in U$ the germs of the functions $f_1, \ldots, f_r$ generate an ideal $\mathfrak{M}_a \subseteq {}_n\mathcal{O}_a$, and it is clear that the set of these ideals form a coherent sheaf of ideals $\mathfrak{M}$ over the open set U; the germ of the function d^s also generates an ideal $\mathfrak{D}_a \subseteq {}_n\mathcal{O}_a$, and again the set of these ideals form a coherent sheaf of ideals $\mathfrak{D}$ over the open set U. The intersection $\mathfrak{M} \cap \mathfrak{D}$ of two coherent sheaves of ideals is also a coherent sheaf of ideals, as a consequence of Oka's theorem; hence, perhaps after shrinking the neighborhood U, there will exist a finite number of analytic functions $h_1, \ldots, h_t \in {}_n\mathcal{O}_U$ such that at each point $a \in U$ the germs of these functions generate the ideal $\mathfrak{M}_a \cap \mathfrak{D}_a \subseteq {}_n\mathcal{O}_a$. Since $h_i \in \mathfrak{D}_a$ for each point $a \in U$, it follows that this function can be written as a product $h_i = d^s h_i'$ for some analytic function $h_i' \in {}_n\mathcal{O}_U$. At the origin the ideal $\mathfrak{M}_o = \text{id}\, V \subset {}_n\mathcal{O}_o$ is prime, by hypothesis; so since $h_i \in \mathfrak{M}_o$ while $d^s \notin \mathfrak{M}_o$, necessarily $h_i' \in \mathfrak{M}_o$. That is to say, there will exist germs $g_{ij} \in {}_n\mathcal{O}_o$ such that $h_i' = \Sigma_j g_{ij} f_j$ hence such that $h_i = \Sigma_j d^s g_{ij} f_j$. Upon restricting the neighborhood U still further if necessary, it can be assumed that the functions g_{ij} are analytic throughout U, and that the last equation above holds in that entire neighborhood. Now to conclude the proof, consider a germ $g \in \text{id}\, V \subseteq {}_n\mathcal{O}_a$ at some point $a \in U$. As noted in the preceding paragraph, $d^s g \in \mathfrak{M}_a$; hence of course $d^s g \in \mathfrak{M}_a \cap \mathfrak{D}_a$. Therefore there are germs $k_i \in {}_n\mathcal{O}_a$ such that $d^s g = \Sigma_i k_i h_i = \Sigma_{ij} d^s k_i g_{ij} f_j$; and dividing through by d^s, as is possible since ${}_n\mathcal{O}_a$ is an integral domain, it follows that

$g = \Sigma_{ij} k_i g_{ij} f_j \in \mathfrak{M}_a$. Therefore $\text{id } V \subseteq \mathfrak{M}_a$ at each point $a \in U$; and since clearly $\mathfrak{M}_a \subseteq \text{id } V$, it follows that $\text{id } V = \mathfrak{M}_a$, thereby completing the proof of the desired result.

It should be observed that the conclusions of Theorem 7 carry over immediately to not necessarily irreducible analytic subvarieties; for if in some open set U the analytic subvariety V can be written $V = V' \cup V''$ and there are analytic functions f'_i , f''_j which generate the ideals of V' and V'' respectively at each point of U , then the products $f'_i \cdot f''_j$ generate the ideal of V at each point of U , so that the obvious induction can be carried out.

If V is any analytic subvariety of an open set $U \subseteq \mathbb{C}^n$, then to each point $a \in U$ there can be associated the ideal $\mathfrak{M}_a = \text{id } V \subseteq {}_n\mathcal{O}_a$; if $a \notin V$ this associated ideal is of course the trivial ideal $\mathfrak{M}_a = {}_n\mathcal{O}_a$. The set of all these ideals form an analytic subsheaf of the sheaf ${}_n\mathcal{O}$ over the set U , which will be denoted by $\mathcal{J}(V)$ and called the <u>sheaf of ideals of the analytic subvariety</u> V . As Theorem 7 and the remarks in the last paragraph show, this is a finitely generated subsheaf of the sheaf ${}_n\mathcal{O}$, hence must be a coherent analytic sheaf over the set U . That is to say, an immediate consequence of Theorem 7 is the following:

<u>Corollary to Theorem 7.</u> If V is any analytic subvariety of an open set $U \subseteq \mathbb{C}^n$, its sheaf of ideals $\mathcal{J}(V)$ is a coherent analytic subsheaf of ${}_n\mathcal{O}$ over U .

(c) It is perhaps worthwhile summarizing some convenient criteria that a system of coordinates be regular for an ideal, so that the local parametrization theorem can be applied to the locus of that ideal.

<u>Theorem 8(a).</u> Suppose that $\mathfrak{A}$ is an ideal in ${}_n\mathcal{O}$, and that V is an analytic subvariety of an open neighborhood of the origin in $\mathbf{C}^n$ representing the germ $\operatorname{loc} \mathfrak{A}$. Then the following three conditions are equivalent:

(i) ${}_{n-1}\mathcal{O}[z_n] \cap \mathfrak{A}$ contains a Weierstrass polynomial in z_n;

(ii) there are arbitrarily small open product neighborhoods $U = U' \times U'' \subseteq \mathbf{C}^{n-1} \times \mathbf{C}$ of the origin in $\mathbf{C}^n$ such that the mapping $\pi: V \cap U \longrightarrow U'$ induced by the natural projection $U' \times U'' \longrightarrow U'$ is a proper, light, continuous mapping;

(iii) the germ of the subvariety $V \cap \{z | z_1 = \dots = z_{n-1} = 0\}$ at the origin is just the origin itself, or equivalently, the origin is an isolated point of this intersection.

Proof. It follows easily as in the proof of Theorem 3 that condition (i) implies condition (ii). Assuming condition (ii), the set $V \cap \{z | z_1 = \dots = z_{n-1} = 0\}$ is just the inverse image of the origin under the light proper mapping π, hence is a finite set of points including the origin; so that condition (ii) implies condition (iii). Assuming condition (iii), note that the function z_n vanishes on the analytic subvariety $V \cap \{z | z_1 = \dots = z_{n-1} = 0\}$ in some open neighborhood of the origin in $\mathbf{C}^n$; hence by the Hilbert zero theorem, $z_n^r = f + g_1 z_1 + \dots + g_{n-1} z_{n-1}$ for some positive integer r and some germs $f \in \mathfrak{M}$, $g_1, \dots, g_{n-1} \in {}_n\mathcal{O}$. The element $f \in \mathfrak{M}$ is thus clearly regular in z_n ; so by the Weierstrass preparation theorem a unit multiple of f will be a Weierstrass polynomial in ${}_{n-1}\mathcal{O}[z_n] \cap \mathfrak{M}$, hence condition (iii) implies condition (i). That suffices to conclude the proof.

Theorem 8(b). Suppose that $\mathfrak{P}$ is a prime ideal in ${}_n\mathcal{O}$, and that V is an analytic subvariety of an open neighborhood of the origin in $\mathbf{C}^n$ representing the germ $\operatorname{loc} \mathfrak{P}$. Then the following three conditions are equivalent:

(i) after a change of coordinates involving only the variables $z_1, \dots, z_k$, these coordinates form a regular system of coordinates for the ideal $\mathfrak{P}$ with respect to which the ideal has dimension at most k ;

(ii) there are arbitrarily small open product neighborhoods $U = U' \times U'' \subseteq \mathbf{C}^k \times \mathbf{C}^{n-k}$ of the origin in $\mathbf{C}^n$ such that the mapping $\pi: V \cap U \longrightarrow U'$ induced by the natural projection mapping $U' \times U'' \longrightarrow U'$ is a proper, light, continuous mapping;

(iii) the germ of the subvariety $V \cap \{z \mid z_1 = \dots = z_k = 0\}$ at the origin is just the origin itself, or equivalently, the origin is an isolated point of this intersection.

Proof. It follows as in the Corollary to Theorem 3 that condition (i) implies condition (ii). Assuming condition (ii), the set $V \cap \{z | z_1 = \cdots = z_k = 0\}$ is just the inverse image of the origin under the light proper mapping π, hence is a finite set of points including the origin; so that condition (ii) implies condition (iii). Assuming condition (iii), note that the germ of the subvariety $V \cap \{z | z_1 = \cdots = z_{n-1} = 0\}$ at the origin is just the origin itself; hence from Theorem 8(a) it follows that ${}_{n-1}\mathcal{O}[z_n] \cap \mathcal{J}$ contains a Weierstrass polynomial in z_n. This implies that z_n is part of a regular system of coordinates for the ideal $\mathcal{J}$, with respect to which the ideal has dimension at most $n-1$, although of course the coordinates $z_1, \ldots, z_{n-1}$ might have to be changed. At any rate, the local parametrization theorem (in particular Corollary 6 to Theorem 5) shows that the natural projection $\mathbb{C}^{n-1} \times \mathbb{C} \longrightarrow \mathbb{C}^{n-1}$ induces a light proper mapping from V onto an analytic subvariety V_{n-1} of an open neighborhood of the origin in $\mathbb{C}^{n-1}$. Now the germ of the subvariety $V_{n-1} \cap \{z | z_1 = \cdots = z_{n-2} = 0\}$ at the origin also necessarily just the origin itself, provided that $k \leq n-2$, so the argument can be repeated with the subvariety V_{n-1} in place of V. The obvious induction shows then that condition (iii) implies condition (i), and the proof is thereby concluded.

It is evident that, when the three equivalent conditions of Theorem 8(b) hold, the coordinates $z_1, \ldots, z_n$ form a regular system of coordinates for the ideal $\mathcal{J}$ with respect to which that ideal has dimension exactly equal to k provided that either (i) ${}_k\mathcal{O} \cap \mathcal{J} = 0$

or (ii) the image of V under the natural projection to U' is all of U' . With this remark it is apparent that Theorem 8(b) contains the converse of Theorem 3, at least for prime ideals.

Theorem 8(c). Suppose that $\mathcal{Y}$ is a prime ideal in ${}_n\mathcal{O}$, that V is an analytic subvariety of an open neighborhood of the origin in $\mathbb{C}^n$ representing the germ loc $\mathcal{Y}$, and that $z_1,\dots,z_n$ form a regular system of coordinates for the ideal $\mathcal{Y}$ with respect to which the ideal has dimension k . Then $z_1,\dots,z_n$ form a strictly regular system of coordinates for the ideal $\mathcal{Y}$ if and only if for sufficiently small open product neighborhoods $U = U' \times U'' \subseteq \mathbb{C}^{k+1} \times \mathbb{C}^{n-k-1}$ of the origin in $\mathbb{C}^n$, the mapping $V \cap U \longrightarrow \pi_{k+1}(V) \cap U'$ induced by the natural projection mapping $U' \times U'' \longrightarrow U'$ is a one-to-one mapping from a dense open subset of $V \cap U$ onto a dense open subset of $\pi_{k+1}(V) \cap U'$, where $\pi_{k+1}(V)$ is the partial projection of V into $\mathbb{C}^{k+1}$.

Proof. If $z_1,\dots,z_n$ do form a strictly regular system of coordinates for the ideal $\mathcal{Y}$, the desired result is an immediate consequence of Theorem 5. For the converse direction, recall from Corollary 6 to Theorem 5 that the partial projection of loc $\mathcal{Y}$ into $\mathbb{C}^{k+1}$ is an irreducible germ of a proper analytic subvariety at the origin in $\mathbb{C}^{k+1}$, and that in a suitable open neighborhood of the origin the natural projection mapping from $\mathbb{C}^n$ to $\mathbb{C}^k$ exhibits dense open subsets of both V and $\pi_{k+1}(V)$ as covering spaces of the complement of an analytic subvariety of an open neighborhood of the origin in $\mathbb{C}^k$; the hypotheses further imply that these are covering spaces of the same number r of sheets. As in the proof of the last part of Theorem 5, every element $\tilde{f}$ of the quotient field ${}_n\tilde{\mathfrak{M}}$ of the residue class ring ${}_n\mathcal{O}/\mathfrak{P}$ is of degree at most r over the subfield ${}_k\tilde{\mathfrak{M}} \cong {}_k\mathfrak{M}$; so that ${}_n\tilde{\mathfrak{M}}$ is an extension field of degree at most r over ${}_k\tilde{\mathfrak{M}} \cong {}_k\mathfrak{M}$. If the degree of the element $\tilde{z}_{k+1} \in {}_n\tilde{\mathfrak{M}}$ over ${}_k\tilde{\mathfrak{M}} \cong {}_k\mathfrak{M}$ is less than r , there must be a monic polynomial $p_{k+1} \in {}_k\mathcal{O}[z_{k+1}] \cap \mathfrak{P}$ of degree less than r ; but since p_{k+1} vanishes on the partial projection $\pi_{k+1}(V)$, that set must be a covering of fewer than r sheets over $\mathbf{C}^k$, which is impossible. Thus $\tilde{z}_{k+1}$ is of degree r , hence generates the field extension ${}_n\tilde{\mathfrak{M}}$ over ${}_k\tilde{\mathfrak{M}} \cong {}_k\mathfrak{M}$; and the given coordinate system is strictly regular for the ideal $\mathfrak{P}$, as desired.

(d) In the definition of a regular system of coordinates for an ideal $\mathfrak{A} \subset {}_n\mathcal{O}$, the notion of the dimension of the ideal with respect to that system of coordinates was introduced; in general this dimension depends both on the ideal and on the choice of the coordinate system. However, if there is a regular system of coordinates for a prime ideal $\mathfrak{P} \subset {}_n\mathcal{O}$ with respect to which the ideal has dimension k , it follows from the local parametrization theorem that a dense open subset of a sufficiently small neighborhood of the origin in any analytic subvariety representing the germ $\operatorname{loc}\mathfrak{P}$, is a k-dimensional complex analytic manifold; so it is clear that for a prime ideal, this dimension is independent of the choice of the coordinate system. Thus it is possible to speak simply of the <u>dimension</u>

of a prime ideal $\mathfrak{p} \subset {}_n\mathcal{O}$, denoting this by $\dim \mathfrak{p}$. For an irreducible germ V of an analytic subvariety, the dimension of the germ V will be defined as the dimension of the ideal $\text{id}\, V$, and will be denoted by $\dim V$. For an arbitrary germ V of an analytic subvariety, written as the union of its irreducible components $V = \cup_i V_i$, the dimension of the germ V will be defined by $\dim V = \max_i \dim V_i$; of course this does not necessarily coincide with the dimension of the ideal $\text{id}\, V$ with respect to all regular systems of coordinates for that ideal. The germ V will be called pure dimensional if $\dim V = \dim V_i$ for all the components V_i.

Several properties of the dimension follow quite readily from the local parametrization theorem, and will be gathered together in the following theorem.

Theorem 9(a). If V and W are germs of analytic subvarieties at the origin in $\mathbb{C}^n$ such that V is irreducible and $W \subset V$, then $\dim W < \dim V$.

Proof. For the proof it can of course be assumed that W is irreducible. Choose a strictly regular system of coordinates for the prime ideal $\text{id}\, V \subset {}_n\mathcal{O}$, with respect to which that ideal has dimension k; since $\text{id}\, V \subset \text{id}\, W$, it is clear that these coordinates can also be taken to be a regular system of coordinates for the prime ideal $\text{id}\, W$, with respect to which that ideal has dimension $\leq k$. Suppose that actually $\dim W = \dim V = k$. Then, by the local parametrization theorem, under the natural projection mapping representative subvarieties for both V and W in some open neighborhood of the origin in $\mathbb{C}^n$ appear as finite-sheeted branched covering spaces of an open neighborhood U' of the origin in $\mathbb{C}^k$; the unbranched part V-B of the covering V is a k-dimensional complex analytic manifold, and an open subset of any open neighborhood of the origin in this manifold is necessarily contained in the subset W . Now any analytic function f in this neighborhood of the origin in $\mathbb{C}^n$, representing a germ $f \in \text{id } W$, restricts to an analytic function on the complex manifold V-B , which vanishes in an open subset of that manifold; and since V-B is connected for an irreducible subvariety V , this function vanishes identically on V-B , hence represents a germ $f \in \text{id } V$. This implies that $\text{id } W \subseteq \text{id } V$, which is impossible since $\text{id } V \subset \text{id } W$; and therefore necessarily $\dim W < \dim V$, as desired.

Theorem 9(b). The germ V of an analytic subvariety at the origin in $\mathbb{C}^n$ has dimension $\leq k$ if and only if for some system of coordinates $z_1, \ldots, z_n$ centered at the origin the germ of the subvariety $V \cap \{z | z_1 = \ldots = z_k = 0\}$ at the origin is just the origin itself.

Proof. This is an immediate consequence of Theorem 8(b).

Theorem 9(c). The germ V of an analytic subvariety at the origin in $\mathbb{C}^n$ is of pure dimension n-1 if and only if id V is a principal ideal.

Proof. First suppose that V is an irreducible germ of an analytic subvariety at the origin in $\mathbb{C}^n$, of dimension n-1 ; and choose a strictly regular coordinate system for the prime ideal id V

In this case there is but a single canonical polynomial, the irreducible Weierstrass polynomial p_n ; so the canonical ideal $\mathcal{K}$ is the prime principal ideal generated by that Weierstrass polynomial. Since $\mathcal{K}$ is a prime ideal, it follows immediately from Theorem 4 that $\mathcal{K} = \text{id } V$, and hence $\text{id } V$ is also a principal ideal, as desired. More generally, if V is an arbitrary germ of pure dimension $n-1$, then writing this germ as the union of its irreducible components $V = V_1 \cup \ldots \cup V_r$, each germ V_i is of dimension $n-1$; hence as above, each ideal $\text{id } V_i$ is a principal ideal, generated by some element $f_i \in {}_n\mathcal{O}$. Since ${}_n\mathcal{O}$ is a unique factorization ring, it is evident that $\text{id } V = \text{id } V_1 \cap \ldots \cap \text{id } V_r$ is the principal ideal generated by the element $f_1 \cdots f_r \in {}_n\mathcal{O}$.

Conversely, suppose that V is the germ of an analytic subvariety at the origin in $\mathbf{C}^n$ such that $\text{id } V$ is the principal ideal generated by some element $f \in {}_n\mathcal{O}$. Since ${}_n\mathcal{O}$ is a unique factorization ring, this element f can be written as a product $f = f_1 \cdots f_r$ of irreducible elements $f_i \in {}_n\mathcal{O}$; and then $V = \text{loc}({}_n\mathcal{O} f) = \text{loc}({}_n\mathcal{O} f_1) \cup \ldots \cup \text{loc}({}_n\mathcal{O} f_r)$, where $V_i = \text{loc } {}_n\mathcal{O} f_i$ are the irreducible components of the germ V since the ideals ${}_n\mathcal{O} f_i$ are prime ideals. Considering any one component V_i , after a suitable choice of coordinates in $\mathbf{C}^n$ it can be assumed that f_i is a Weierstrass polynomial in z_n . If $\dim V_i < n-1$, there is necessarily an element $h \in \text{id } V_i = {}_n\mathcal{O} f_i$ independent of z_n ; that is to say, for some nonzero element $g \in {}_n\mathcal{O}$ the product $h = g f_i$ is independent of z_n . However, for any point $z' = (z_1, \ldots, z_{n-1})$ sufficiently near the origin such that $h(z') \neq 0$, there will be some value of z_n in the region of analyticity of representatives of all these germs such that $f_i(z_1,\ldots,z_n) = 0$; so this is clearly impossible, showing that $\dim V_i = n-1$ and concluding the proof.

Theorem 9(d). The germ V of an analytic subvariety at the origin in $\mathbb{C}^n$ is of dimension 0 if and only if V consists of the origin itself; this is the only germ that can be represented by a compact analytic subvariety of an open neighborhood of the origin in $\mathbb{C}^n$.

Proof. The first part is an immediate consequence of the local parametrization theorem. For the second part, it suffices to show that if V is a connected complex analytic subvariety of an open subset $U \subseteq \mathbb{C}^n$, the restriction to V of any analytic function in U cannot attain its maximum modulus unless it is constant on V; for if V is a compact connected analytic subvariety of U, then all functions analytic in U, and in particular the coordinate functions in $\mathbb{C}^n$, are necessarily constant on V, and hence V consists of a single point. Suppose then that the restriction to V of an analytic function f in U attains its maximum at a point $p \in V$; it can of course be assumed that p is the origin in $\mathbb{C}^n$, that V is an irreducible germ at the origin, and that V is represented as an r-sheeted branched covering of an open neighborhood of the origin in $\mathbb{C}^k$, as in the local parametrization theorem. To the function f there is associated a monic polynomial $p_f(z';X) \in {}_k\mathcal{O}[X]$ such that $p_f(z';f(z)) = 0$ whenever $z = (z',z'') \in V$

The values of the coefficients of this polynomial at any point $z' \in \mathbb{C}^k$ sufficiently near the origin are the elementary symmetric functions of the values taken on by the function f at the r points $z_j = (z', z''_j) \in V$ lying over the point z' ; and these are complex analytic functions of z' . However, since f attains its maximum at the single point lying over $z' = 0$, it follows easily from the usual maximum modulus theorem that these coefficients must indeed be constant; but then the function f must itself be constant on the subvariety V , as desired.

Before continuing with further parts of the theorem, it is convenient to demonstrate the following useful auxiliary result.

<u>Semicontinuity Lemma.</u> Suppose that $f_1(z;t), \ldots, f_r(z;t)$ are continuous functions in an open subset $U' \times U'' \subseteq \mathbb{C}^n \times \mathbb{C}^m$, and are analytic in $z \in U'$ for each fixed $t \in U''$; and for each fixed $t \in U''$ let $V(t)$ be the germ at the origin of the analytic subvariety $\{z \in U' | f_1(z;t) = \ldots = f_r(z;t) = 0\}$, the origin being a point of U' . Then for any fixed $t_o \in U''$ it follows that $\dim V(t) \leqq \dim V(t_o)$ whenever t is sufficiently near t_o . (The function $\dim V(t)$ is thus an upper semicontinuous function of t in U'' .)

Proof. If $\dim V(t_o) = k$, it is a consequence of Theorem 9(b) that for a suitable system of coordinates in U' the origin is an isolated point of the subvariety $V(t_o) \cap \{z \in U' | z_1 = \ldots = z_k = 0\}$; consequently for some positive numbers δ, ε it follows that

$$\sum_i |f_i(0,\ldots,0,z_{k+1},\ldots,z_n;t_o)| \geqq \varepsilon > 0 \quad \text{whenever} \quad \max_{k+1 \leq j \leq n} |z_j| = \delta .$$

By continuity then,

$$\sum_i |f_i(0,\ldots,0,z_{k+1},\ldots,z_n;t)| \geq \frac{\varepsilon}{2} > 0 \quad \text{whenever} \quad \max_{k+1 \leq j \leq n} |z_j| = \delta$$

for all points t sufficiently near t_o. The subvariety of $U'' \subseteq \mathbf{C}^{n-k}$ defined by the equations $f_i(0,\ldots,0,z_{k+1},\ldots,z_n;t) = 0$ for $i = 1,\ldots,r$ is thus disjoint from the boundary of the polydisc of radius δ centered at the origin in $\mathbf{C}^{n-k}$; so the component of that subvariety contained in the interior of the polydisc is necessarily compact, hence is of dimension 0 by Theorem 9(d). That is to say, $V(t) \cap \{z \in U' | z_1 = \ldots = z_k = 0\}$ is either empty or has the origin as an isolated point, so that by Theorem 9(b) again $\dim V(t) \leq k$, as desired.

Theorem 9(e). If V is the germ at the origin in $\mathbf{C}^n$ of an analytic subvariety of pure dimension k, and if $f \in {}_n\mathcal{O}$ is a non-unit which does not vanish identically on any irreducible component of V, then the germ at the origin of the subvariety $W = V \cap \{z | f(z) = 0\}$ is of pure dimension $k-1$.

Proof. Of course it suffices to prove this theorem for the special case that V is an irreducible germ. Since $f \notin \text{id}\, V$, it follows that $W \subset V$, and hence by Theorem 9(a) that $\dim W < \dim V = k$. Let $f_1,\ldots,f_r$ be analytic functions in an open neighborhood of the origin defining a subvariety representing the germ V; and for any point $t \in V$ sufficiently near the origin, consider the germ $W(t)$ at the origin of the analytic subvariety $\{z | f_1(z+t) = \ldots = f_r(z+t) = f(z) = 0\}$. Note that this germ is analytically equivalent to the germ at the point t of the subvariety $\{z|f_1(z) = \ldots = f_r(z) = f(z-t) = 0\} = V \cap \{z|f(z-t) = 0\}$; so if $t \in V$ is any point at which V is a k-dimensional complex analytic manifold, then since $f(z-t)$ vanishes at that point it follows from Theorem 9(c) that $\dim W(t) = k-1$. Note also that $W(0) = W$; and since there are points $t \in V$ arbitrarily near the origin at which V is a k-dimensional complex analytic manifold, it follows from the semicontinuity lemma that $k-1 = \dim W(t) \leqq \dim W$. Therefore it is at least true that $\dim W = k-1$. If W is irreducible, it is then necessarily of pure dimension $k-1$. If W is reducible, write it as the union $W = W_1 \cup \ldots \cup W_s$ of its irreducible components, and let $k_i = \dim W_i$. For each component W_i select a point $a_i \in W_i$ such that W_i is a k_i-dimensional complex analytic manifold near a_i and such that $a_i \notin W_j$ for $j \neq i$. Near the point $a_i \in V$ of course W_i is irreducible and is defined by the vanishing of the single analytic function f ; so by what has just been proved, necessarily $k_i = k-1$, and the desired result is thereby demonstrated.

Some comments should perhaps be made at this point about the preceding theorem. First, if V is an analytic subvariety in an open neighborhood of the origin, and if V is irreducible at the origin, it is not necessarily irreducible at all points near the origin; for example, the subvariety $V \subset \mathbb{C}^3$ defined by $V = \{(z_1,z_2,z_3) \in \mathbb{C}^3 | z_1^2 - z_2^2 z_3 = 0\}$ is readily seen to be irreducible at the origin, but to be reducible at any point $(0,0,z_3)$ for $z_3 \neq 0$. However, V is at least pure dimensional at all points sufficiently near the origin; for if V is irreducible and of dimension k at the origin, a dense open subset of an open neighborhood of the origin in V is a k-dimensional complex analytic manifold. This observation was used in the proof of Theorem 9(e), as the reader will no doubt have observed. Second, if V is a complex analytic manifold, then the converse to Theorem 9(e) also holds; indeed, this is really just Theorem 9(c). However, for a general analytic subvariety $V \subset \mathbb{C}^n$, the converse to Theorem 9(e) does not necessarily hold; that is to say, if V is of pure dimension k and if $W \subset V$ is an analytic subvariety of pure dimension $k-1$, it is not necessarily true that $W = V \cap \{z | f(z) = 0\}$ for some analytic function f. This is a point to which further discussion will later be given. Third, by iterating Theorem 9(e) in the obvious manner, it follows readily that whenever $f_1, \ldots, f_{n-k}$ are analytic functions in an open neighborhood U of the origin in $\mathbb{C}^n$, the subvariety $V = \{z \in U | f_1(z) = \ldots = f_{n-k}(z) = 0\}$ has dimension at least k at each of its points. Again however, as might be expected in view of the preceding comments, the converse assertion does not necessarily hold; a subvariety of pure dimension k in $\mathbb{C}^n$ cannot necessarily be defined as the set of common zeros of precisely $n-k$ analytic functions, even locally.

Theorem 9(f). If V_1, V_2 are germs at the origin in $\mathbb{C}^n$ of analytic subvarieties of pure dimensions k_1, k_2, respectively, then for any component W of the intersection $V_1 \cap V_2$ it follows that $\dim W \geqq k_1 + k_2 - n$.

Proof. Let $f_1,\dots,f_r,g_1,\dots,g_s$ be analytic functions in an open neighborhood of the origin such that the subvarieties $V_1 = \{z|f_1(z)=\dots=f_r(z)=0\}$ and $V_2=\{z|g_1(z)=\dots=g_s(z)=0\}$ represent the germs V_1, V_2, respectively. For any point $t \in V_1$ sufficiently near the origin consider the germ $W(t)$ at the origin of the analytic subvariety

$\{z|f_1(z+t) = \dots = f_r(z+t) = g_1(z) = \dots = g_s(z) = 0\}$. Note that this is just the intersection of the germ V_2 with the translation to the origin of the germ of the subvariety V_1 at the point $t \in V_1$. If $t \in V_1$ is any point at which V_1 is a k_1-dimensional complex analytic manifold, then $\dim W(t) \geq k_1 + k_2 - n$; for at such a point the submanifold V_1 can be defined by the vanishing of $n-k_1$ coordinate functions in $\mathbf{C}^n$, hence it follows readily from Theorem 9(e) that for the subset $W(t) \subseteq V_2$ defined by these functions necessarily $\dim W(t) \geqq k_2 - (n-k_1)$. Note further that $W(0) = V_1 \cap V_2$; and since there are points $t \in V_1$ arbitrarily near the origin at which V_1 is a k_1-dimensional manifold, it follows from the semicontinuity lemma that $\dim V_1 \geqq \dim W(t) \geq k_1 + k_2 - n$. If $V_1 \cap V_2$ is irreducible, the desired result has been demonstrated. If $V_1 \cap V_2$ is reducible, write it as the union $W = W_1 \cup \dots \cup W_m$ of its irreducible components, and let $\ell_i = \dim W_i$. For each component W_i select a point $a_i \in W_i$ such that W_i is a ℓ_i-dimensional complex analytic manifold near a_i and such that $a_i \notin W_j$ for $j \neq i$. Near the point a_i of course W_i is irreducible and $W_i = V_1 \cap V_2$; so by what has just been proved, $\ell_i \geqq k_1 + k_2 - n$, and the theorem is thus proved.

§4. Analytic varieties and their local rings

(a) In the discussion of the local parametrization theorem, interest was centered on the form of an analytic subvariety in terms of a particular, conveniently chosen system of coordinates in the ambient space $\mathbb{C}^n$. In the applications of the local parametrization theorem discussed in the last section, however, the role of a particular coordinate system was irrelevant, except as a tool in the derivation of the desired properties. For these and many other properties interest really lies in an equivalence class of germs of analytic subvarieties, where two germs V_1, V_2 of analytic subvarieties at the origin in $\mathbb{C}^n$ are called equivalent germs of analytic subvarieties if there are representative subvarieties V_1, V_2 in open neighborhoods U_1, U_2 of the origin in $\mathbb{C}^n$ and an analytic homeomorphism $\varphi: U_1 \longrightarrow U_2$ such that $\varphi(V_1) = V_2$. It is obvious that this is indeed an equivalence relation in the technical sense. There will generally be no attempt made to distinguish between germs of analytic subvarieties and equivalence classes of germs of analytic subvarieties; it is usually completely clear from context which is meant.

It should be noted that an equivalence class of germs of analytic subvarieties depends quite essentially on the particular imbeddings of the subvarieties in the ambient space $\mathbb{C}^n$. Thus the germ of a k-dimensional analytic submanifold of $\mathbb{C}^{n_1}$ and the germ of a k-dimensional analytic submanifold of $\mathbb{C}^{n_2}$ are inequivalent germs of analytic subvarieties whenever $n_1 \neq n_2$, even though they are equivalent germs of complex analytic manifolds; and again, if V is the germ of an analytic subvariety at the origin in $\mathbb{C}^n$, then V can also be viewed as the germ of an analytic subvariety at the origin in $\mathbb{C}^{n+1}$ through the canonical imbedding $\mathbb{C}^n \subset \mathbb{C}^{n+1}$, but these are inequivalent germs of analytic subvarieties. It is thus evident that there is a point to introducing a further, weaker equivalence relation among germs of analytic subvarieties, in order to investigate those properties of analytic subvarieties which are to some extent independent of the imbeddings of these subvarieties in their ambient complex number spaces.

For this purpose, consider two germs V_1, V_2 of analytic subvarieties at the origin in spaces $\mathbb{C}^{n_1}, \mathbb{C}^{n_2}$, respectively. By a <u>continuous mapping</u> from the germ V_1 into the germ V_2 is meant the germ at the origin of a continuous mapping $\varphi\colon V_1 \rightarrow V_2$, where here V_1 and V_2 are analytic subvarieties in some open neighborhoods U_1, U_2 of the origin in the spaces $\mathbb{C}^{n_1}, \mathbb{C}^{n_2}$ respectively, representing the given germs. The two germs V_1, V_2 are <u>topologically equivalent</u> if there are continuous mappings $\varphi\colon V_1 \rightarrow V_2$ and $\psi\colon V_2 \rightarrow V_1$ such that the compositions $\psi\varphi\colon V_1 \rightarrow V_1$ and $\varphi\psi\colon V_2 \rightarrow V_2$ are the identity maps; this is of course equivalent to the condition that the two germs have topologically homeomorphic representative subvarieties in some open neighborhoods of the origin. A continuous mapping $\varphi\colon V_1 \rightarrow V_2$ is said to be an <u>analytic mapping</u> from the germ V_1 into the germ V_2 if a representative mapping on analytic subvarieties extends to an analytic mapping of a neighborhood of the origin in $\mathbb{C}^{n_1}$ into a neighborhood of the origin in $\mathbb{C}^{n_2}$; that is to say, the mapping $\varphi\colon V_1 \longrightarrow V_2$ is analytic if in terms of some representative subvarieties V_1, V_2 in open neighborhoods U_1, U_2 of the origin in the spaces $\mathbb{C}^{n_1}$, $\mathbb{C}^{n_2}$, respectively, there is an analytic mapping $\Phi\colon U_1 \longrightarrow U_2$ such that $\Phi|V_1 = \varphi$. Note that the critical matter is that there exists some extension of φ to an analytic mapping, but not what the particular extension is; so two analytic mappings are identified when they yield the same continuous mapping from V_1 into V_2, regardless of what the extensions of these mappings are in the ambient complex number spaces. The two germs V_1, V_2 are said to determine <u>equivalent germs of analytic varieties</u> if there are analytic mappings $\varphi\colon V_1 \longrightarrow V_2$ and $\psi\colon V_2 \longrightarrow V_1$ such that the compositions $\psi\varphi\colon V_1 \longrightarrow V_1$ and $\varphi\psi\colon V_2 \longrightarrow V_2$ are the identity maps. It is clear that this is an equivalence relation in the technical sense; an equivalence class is called <u>the germ of an analytic variety</u>.

Note that equivalent subvarieties in this sense are topologically equivalent spaces; so that underlying any germ of an analytic variety is a well defined germ of a topological space. Note further however that a germ of an analytic variety cannot be viewed as being imbedded in a complex number space $\mathbb{C}^n$, although of course a representative analytic subvariety is always so imbedded; different representatives of the same variety may be imbedded in quite different complex number spaces.

(b) The germ of an analytic variety can be viewed as the germ of a topological space with an additional structure imposed; and one way of describing this additional structure is through the analytic functions on the variety.

Consider first an analytic subvariety V of an open subset $U \subseteq \mathbb{C}^n$; and to each point $a \in V$ associate the ideal $\mathfrak{A}_a = \text{id}\, V \subset {}_n\mathcal{O}_a$. The residue class ring ${}_n\mathcal{O}_a/\mathfrak{A}_a$, which was considered in some detail earlier, will now be denoted by ${}_V\mathcal{O}_a$, and will be called <u>the ring of germs of holomorphic functions on the subvariety</u> V <u>at the point</u> a . The terminology is suggested by the following outlook on this residue class ring. For any germ $f \in {}_n\mathcal{O}_a$ select a representative analytic function f in an open neighborhood of the point a in $\mathbb{C}^n$; the restriction of this function to the subvariety V is a continuous complex-valued function in an open neighborhood of the point a on the set V , and the germ of this restricted function at the point a clearly depends only on the original germ $f \in {}_n\mathcal{O}_a$. It is apparent that this restriction mapping is a well-defined homomorphism from the ring ${}_n\mathcal{O}_a$ into the ring of germs of continuous complex-valued functions on the set V at the point a , and that the kernel of this homomorphism is the ideal $\mathfrak{A}_a$; hence the residue class ring ${}_V\mathcal{O}_a$ can be identified with a subring of the ring of germs of continuous complex-valued functions on the set V at the point a , and the germs so arising can be considered to be the germs of holomorphic functions on the subvariety V at the point a .

The set of rings ${}_V\mathcal{O}_a$ for all points $a \in V$ can be taken to form a sheaf of rings over V which will be denoted by ${}_V\mathcal{O}$ and called the sheaf of germs of holomorphic functions on the subvariety V. Note that this sheaf can be viewed as a subsheaf of rings in the sheaf of germs of continuous complex-valued functions on the set V. For any relatively open subset $W \subseteq V$ the ring $\Gamma(W, {}_V\mathcal{O})$ of sections of the sheaf ${}_V\mathcal{O}$ over W will also be denoted by ${}_V\mathcal{O}_W$ and will be called the ring of holomorphic functions in the subset W of the subvariety V. Any section $f \in \Gamma(W, {}_V\mathcal{O}) = {}_V\mathcal{O}_W$ can of course be viewed as a continuous complex-valued function on the set W; and a continuous complex-valued function f on the set W belongs to the ring $\Gamma(W, {}_V\mathcal{O}) = {}_V\mathcal{O}_W$ if and only if it is locally the restriction to V of an analytic function in the ambient space $\mathbb{C}^n$. It should be emphasized that it is not required that there should exist an analytic function F in an open neighborhood of W in $\mathbb{C}^n$ such that $F|W = f$; it is only required that this should be true locally. Note that the restriction homomorphism ${}_n\mathcal{O}_a \longrightarrow {}_V\mathcal{O}_a$ exhibits ${}_V\mathcal{O}_a$ as a module over the ring ${}_n\mathcal{O}_a$; hence ${}_V\mathcal{O}$ can be viewed as a sheaf of modules over the sheaf of rings ${}_n\mathcal{O}|V$ on the set V. Actually of course when viewed in this light ${}_V\mathcal{O}$ is just the restriction to the subvariety V of the analytic sheaf $({}_n\mathcal{O}|U)/\mathcal{J}(V)$, where as before $\mathcal{J}(V) \subset {}_n\mathcal{O}|U$ is the sheaf of ideals of the subvariety V.

Now consider two germs V_1, V_2 of analytic subvarieties at the origin in spaces $\mathbb{C}^{n_1}$, $\mathbb{C}^{n_2}$ respectively. If there is a continuous mapping $\varphi: V_1 \longrightarrow V_2$, then for each germ f of a continuous complex-valued function on V_2 the composition $f\circ\varphi$ is a well-defined germ of a continuous complex-valued function on V_1; thus φ induces a homomorphism φ^* from the ring of germs of continuous complex-valued functions on V_2 into the ring of germs of continuous complex-valued functions on V_1. In particular, φ^* maps the ring ${}_{V_2}\mathcal{O}$ into the ring of germs of continuous complex-valued functions on V_1.

<u>Theorem 10.</u> A continuous mapping $\varphi\colon V_1 \longrightarrow V_2$ between two germs of analytic subvarieties is analytic if and only if $\varphi^*({}_{V_2}\mathcal{O}) \subseteq {}_{V_1}\mathcal{O}$.

Proof. Select analytic subvarieties V_1, V_2 of open neighborhoods U_1, U_2 of the origin, representing the given germs of analytic subvarieties, and a continuous mapping $\varphi\colon V_1 \longrightarrow V_2$ representing the given germ of a continuous mapping. If φ is analytic then, perhaps after shrinking the neighborhoods, there is a complex analytic mapping $\Phi\colon U_1 \longrightarrow U_2$ such that $\Phi|V_1 = \varphi$. Now a germ $f \in {}_{V_2}\mathcal{O}$ can be represented by the restriction $F|V_2$ for some analytic function F in U_2, again shrinking the neighborhoods if necessary; the germ $\varphi^*(f) = f\circ\varphi$ can then be represented by the restriction $F\circ\Phi|V_1$, and hence $\varphi^*(f) \in {}_{V_1}\mathcal{O}$. Conversely suppose that $\varphi^*({}_{V_2}\mathcal{O}) \subseteq {}_{V_1}\mathcal{O}$. The coordinates $w_1,\ldots,w_{n_2}$ in U_2 restrict to complex analytic functions on the subvariety V_2, and the compositions $\varphi^*(w_j|V_2) = w_j\circ\varphi$ are then analytic functions on the subvariety V_1 near the origin; hence, after shrinking the neighborhoods if necessary, there will exist analytic functions $F_1,\dots,F_{n_2}$ in U_1 such that $F_j|V_1 = w_j\circ\varphi$. These functions F_j can be used as the coordinate functions defining a complex analytic mapping $\Phi: U_1 \longrightarrow U_2$, and it is evident from their construction that $\Phi|V_1 = \varphi$; the mapping φ is thus an analytic mapping, and the proof is thereby completed.

A germ of an analytic subvariety determines a germs of a topological space; and this space further possesses a distinguished subring of the ring of germs of continuous complex-valued functions, namely the ring of germs of holomorphic functions on the subvariety. It is an immediate consequence of Theorem 10 that two germs V_1, V_2 of analytic subvarieties determine equivalent germs of varieties if and only if there is a topological homeomorphism $\varphi: V_1 \longrightarrow V_2$ inducing an isomorphism $\varphi^*: {}_{V_2}\mathcal{O} \longrightarrow {}_{V_1}\mathcal{O}$ between the rings of germs of analytic functions on the two subvarieties. Thus the ring ${}_V\mathcal{O}$ on an analytic subvariety V is the complete invariant determining equivalence as varieties; and consequently the germ of an analytic variety can also be defined as an equivalence class of germs of topological spaces endowed with distinguished subrings of the rings of germs of continuous complex-valued functions, equivalence being topological homeomorphism and the induced mapping of functions, such that the class contains the germs of an analytic subvariety with its ring of germs of holomorphic functions.

With this observation in mind, it is an easy matter to introduce the global extension of the germ of an analytic variety.

An analytic variety is a Hausdorff topological space endowed with a distinguished subsheaf ${}_V\mathcal{O}$ of the sheaf of germs of continuous complex-valued functions on V, such that at each point $a \in V$ the germ of V together with the stalk ${}_V\mathcal{O}_a$ is the germ of an analytic variety. The sheaf ${}_V\mathcal{O}$ will be called the sheaf of germs of holomorphic functions on the analytic variety V; or alternatively, the sheaf ${}_V\mathcal{O}$ will be called the structure sheaf of the analytic variety, since it provides a complete description of the structure of the variety. The sections of the structure sheaf ${}_V\mathcal{O}$ over a relatively open subset $W \subseteq V$ will be denoted by ${}_V\mathcal{O}_W$ and will be called holomorphic functions in the subset W of the analytic variety V; these are of course continuous complex-valued functions on the subset W. It should be noted that a sufficiently small open neighborhood of any point on an analytic variety can be represented by an analytic subvariety of an open subset of some complex number space; but the entire variety may not be representable by an analytic subvariety.

(c) Some of the elementary properties of analytic varieties are quite easily established. At any point a on an analytic variety V, the ring ${}_V\mathcal{O}_a$ of germs of holomorphic functions can be represented as the residue class ring ${}_n\mathcal{O}/\mathfrak{A}$ for some ideal $\mathfrak{A} \subset {}_n\mathcal{O}$, or alternatively, recalling the discussion of the local parametrization theorem, as an integral algebraic extension of the ring ${}_k\mathcal{O}$ for some integer $0 \leq k \leq n$. It follows immediately that ${}_V\mathcal{O}_a$

is a <u>Noetherian ring</u>. The units of the ring ${}_V\mathcal{O}_a$ are those germs of analytic functions which are non-zero at the point a; consequently the non-units form the ideal ${}_V\mathfrak{m}_a \subset {}_V\mathcal{O}_a$ consisting of all germs of analytic functions vanishing at the point a. The ring ${}_V\mathcal{O}_a$ is thus a <u>local ring</u>, with <u>maximal ideal</u> ${}_V\mathfrak{m}_a$; and the residue class field ${}_V\mathcal{O}_a/{}_V\mathfrak{m}_a$ is clearly the complex number field $\mathbb{C}$. Since elements of ${}_V\mathcal{O}_a$ can be viewed as germs of continuous complex-valued functions, it is apparent that ${}_V\mathcal{O}_a$ <u>contains no nilpotent elements</u>; that is to say, no power of an element $f \in {}_V\mathcal{O}_a$ is the zero element unless f is itself the zero element.

The germ V of an analytic variety is said to be <u>reducible</u> if it can be written $V = V_1 \cup V_2$ when $V_i \subset V$ are also germs of analytic varieties; and a germ which is not reducible is said to be <u>irreducible</u>. An analytic variety V is said to be reducible or irreducible at a point $a \in V$ according as the germ of that variety at the point a is reducible or irreducible. Note that the germ of an analytic variety is irreducible precisely when the germ of any representative subvariety is irreducible; it then follows from Theorem 1 that the germ of any analytic variety can be written uniquely as an irredundant union of finitely many irreducible germs of analytic varieties. Note that <u>the germ V of an analytic variety is irreducible if and only if the ring ${}_V\mathcal{O}$ is an integral domain</u>; for considering a representative subvariety V with ideal $\mathfrak{A} = \text{id}\, V \subset {}_n\mathcal{O}$, the residue class ring ${}_n\mathcal{O}/\mathfrak{A}$ is an integral domain precisely when $\mathfrak{A}$ is a prime ideal, and as noted earlier,

$\mathfrak{M} = \text{id } V$ is prime precisely when the germ of the subvariety V is irreducible. It should be pointed out that, even when ${}_V\mathcal{O}$ is an integral domain, it is not necessarily a unique factorization domain; further discussion of this point will be deferred to a later portion of these notes.

An <u>analytic subvariety of an analytic variety</u> V is a subset of V which in some open neighborhood of each point of V is the set of common zeros of a finite number of holomorphic functions in that subset of V ; and as usual, there is correspondingly defined the germ of an analytic subvariety of the germ of the variety V at any point. If W is an analytic subvariety of V , then whenever an open subset of V is represented as an analytic subvariety of an open set U in some complex number space $\mathbb{C}^n$, the part of W contained in that subset of V is represented as another analytic subvariety of U , contained in V . There is a natural correspondence associating to each germ W of an analytic subvariety of the germ V of an analytic variety an ideal id $W \subset {}_V\mathcal{O}$, just as in the case of germs of analytic subvarieties at a point in $\mathbb{C}^n$; and there is further a natural correspondence associating to each ideal $\mathfrak{M} \subset {}_V\mathcal{O}$ a germ loc $\mathfrak{M}$ of an analytic subvariety of the germ V of an analytic variety. These correspondences satisfy the quite obvious relations listed on page 10 for the case that $V = \mathbb{C}^n$. Less obvious but still quite easy is the assertion that the Hilbert zero theorem holds for ideals in the ring ${}_V\mathcal{O}$.

Theorem 11. For any germ V of an analytic variety and any ideal $\mathfrak{A} \subset {}_V\mathcal{O}$ it follows that $\text{id loc}\,\mathfrak{A} = \sqrt{\mathfrak{A}}$.

Proof. Represent the germ V by the germ V of an analytic subvariety at the origin in $\mathbb{C}^n$, and let $\mathfrak{V} = \text{id}\, V \subset {}_n\mathcal{O}$; then ${}_V\mathcal{O}$ is the image of the natural ring homomorphism $\rho: {}_n\mathcal{O} \longrightarrow {}_n\mathcal{O}/\mathfrak{V} \cong {}_V\mathcal{O}$, and the kernel of this homomorphism is the ideal $\mathfrak{V} \subset {}_n\mathcal{O}$. For clarity, the locus of an ideal in ${}_n\mathcal{O}$ will be denoted by loc_n , and the locus of an ideal in ${}_V\mathcal{O}$ will be denoted by loc_V ; and correspondingly, the ideal of a subvariety in $\mathbb{C}^n$ will be denoted by id_n , and the ideal of a subvariety in V will be denoted by id_V . Introducing the ideal $\mathfrak{A}' = \rho^{-1}(\mathfrak{A})$ in the ring ${}_n\mathcal{O}$, note first that $\text{loc}_V\,\mathfrak{A} = \text{loc}_n\,\mathfrak{A}'$. For letting $f_1,\ldots,f_r$ be elements of ${}_n\mathcal{O}$ generating the ideal $\mathfrak{V}$ and $g_1,\ldots,g_s$ be elements of ${}_n\mathcal{O}$ whose images $\rho(g_1),\ldots,\rho(g_s)$ generate the ideal $\mathfrak{A}$ in ${}_V\mathcal{O}$, it is evident that $f_1,\ldots,f_r,g_1,\ldots,g_s$ generate the ideal $\mathfrak{A}'$; now a point z in a sufficiently small neighborhood of the origin in $\mathbb{C}^n$ lies in the analytic subvariety representing $\text{loc}_V\,\mathfrak{A}$ precisely when $z \in V$ and $g_1(z) = \ldots = g_s(z) = 0$, hence precisely when $f_1(z) = \ldots = f_r(z) = g_1(z) = \ldots = g_s(z) = 0$ and thus when z lies in the analytic subvariety representing $\text{loc}_n\,\mathfrak{A}'$. Letting $W = \text{loc}_V\,\mathfrak{A} = \text{loc}_n\,\mathfrak{A}'$, note further that $\text{id}_V\, W = \rho(\text{id}_n\, W)$; for $\rho(h) \in \text{id}_V\, W$ for an element $h \in {}_n\mathcal{O}$ precisely when $h|W = 0$, hence precisely when $h \in \text{id}_n\, W$. Therefore, applying the usual Hilbert zero theorem in $\mathbb{C}^n$, it follows that $\text{id}_V\,\text{loc}_V\,\mathfrak{A} = \rho(\text{id}_n\,\text{loc}_n\,\mathfrak{A}') = \rho(\sqrt{\mathfrak{A}'}) = \sqrt{\mathfrak{A}}$, as desired.

A point in an analytic variety V is said to be a <u>regular point</u> of the variety V if the germ of V at that point is equivalent to the germ of the complex number space $\mathbb{C}^k$ of some dimension k; the set of all regular points form the <u>regular locus</u> of V, which will be dnoted by $\mathcal{R}(V)$. It is clear that $\mathcal{R}(V)$ is a complex analytic manifold, although it is neither necessarily connected nor necessarily pure dimensional; and it is clear that $\mathcal{R}(V)$ is an open subset of the variety V. A variety V such that $V = \mathcal{R}(V)$ will be called a <u>regular analytic variety</u>; evidently a regular analytic variety is just a complex analytic manifold itself. The complement $V - \mathcal{R}(V)$ of the regular locus of V is called the <u>singular locus</u> of the variety V, and will be denoted by $\mathcal{S}(V)$; a point in $\mathcal{S}(V)$ is said to be a <u>singular point</u> of the variety V. If the analytic variety V is represented by an analytic subvariety V of an open subset $U \subseteq \mathbb{C}^n$, then <u>the regular points of the variety V are precisely those points at which the representative subvariety $V \subseteq U$ is a complex analytic submanifold.</u> For on the one hand it is completely obvious that any point of the subvariety $V \subseteq U$ at which V is a complex analytic submanifold of U is necessarily a regular point of the variety V. On the other hand, in the neighborhood of any regular point of the analytic variety there are for the representative analytic subvariety $V \subseteq U$ local complex analytic mappings $\varphi: \mathbb{C}^k \longrightarrow V \subseteq U$ and $\psi: V \longrightarrow \mathbb{C}^k$ such that $\psi\varphi: \mathbb{C}^k \longrightarrow \mathbb{C}^k$ is the identity mapping near the origin; but then the mapping $\varphi: \mathbb{C}^k \longrightarrow U$ is necessarily of rank k near the origin, and hence its image, the subvariety $V \subset U$, is locally a complex analytic submanifold of U . This thus provides an equivalent way of describing the regular locus of the variety V , which is useful in deriving such results as the following.

Theorem 12. For any complex analytic variety V the singular locus $\mathcal{S}(V)$ is a proper analytic subvariety of V .

Proof. Since the result is really local in character, it suffices to consider a representative subvariety V of an open neighborhood U of the origin in $\mathbb{C}^n$; and by the preceding remarks, it suffices to show that the set of points of V at which V is not an analytic submanifold of U form a proper analytic subvariety of V . If the variety V is reducible at the origin, so it can be written as a union of subvarieties $V = V_1 \cup V_2$ provided U is small enough, then clearly $\mathcal{S}(V) = \mathcal{S}(V_1) \cup \mathcal{S}(V_2) \cup (V_1 \cap V_2)$; hence it suffices to prove the desired result when the subvariety V is irreducible at the origin. In this case, V is of pure dimension k at each point; and from the local parametrization theorem it follows that $\mathcal{R}(V)$ is a connected, k-dimensional, complex analytic manifold forming a dense open subset of the analytic subvariety V . Now from Theorem 7 it follows that whenever U is sufficiently small, there are holomorphic functions $f_1,\ldots,f_r$ in U which generate the ideal $\text{id}\, V \subset {}_n\mathcal{O}_a$ at each point $a \in V$. It is then easy to see that $\mathcal{S}(V)$ is the subset of V consisting of those points at which the rank of the Jacobian matrix of the functions $f_1,\ldots,f_r$ is strictly less than n-k . For on the one hand, if the rank of this Jacobian matrix at some point $a \in V$ is

m and $m \geq n-k$, there are m of the functions f_i whose common zeros form an analytic submanifold of dimension $n-m \leq k$; and since this submanifold contains V , it necessarily coincides with V , so that $m = n-k$ and V is a k-dimensional analytic submanifold at the point a . On the other hand, if V is a k-dimensional analytic submanifold at a point $a \in V$, there are $n-k$ analytic functions $g_1, \ldots, g_{n-k}$ such that in a neighborhood of a the subvariety V is the set of common zeros of these functions, and that the rank of the Jacobian matrix of these functions at the point a is just $n-k$; but since the functions g_i are contained in the ideal $\text{id}\, V \subset {}_n\mathcal{O}_a$, it follows that they can be written in the form $g_i = \sum_{j=1}^{r} h_{ij} f_j$ for some analytic functions $h_{ij} \in {}_n\mathcal{O}_a$, and hence it is apparent that the rank of the Jacobian matrix of the functions $f_1, \ldots, f_r$ is also at least $n-k$. The singular locus $\mathcal{S}(V)$ is then the subset of V described by the vanishing of all $(n-k) \times (n-k)$ subdeterminants of the Jacobian matrix of the functions $f_1, \ldots, f_r$, and consequently $\mathcal{S}(V)$ is an analytic subvariety of V as desired.

To conclude this catalog of elementary properties of analytic varieties, it should be observed that the machinery of analytic sheaves can be carried over quite readily to analytic varieties. If V is an analytic variety with structure sheaf ${}_V\mathcal{O}$, an <u>analytic sheaf</u> over V is a sheaf of modules over the sheaf of rings ${}_V\mathcal{O}$. Again the easiest examples are the free analytic sheaves ${}_V\mathcal{O} \oplus \ldots \oplus {}_V\mathcal{O} = {}_V\mathcal{O}^r$; and an analytic sheaf which is the homomorphic image of a free analytic sheaf is called a finitely generated analytic sheaf. The critical matter for applying the machinery of analytic sheaves is of course Oka's theorem, which does extend quite easily to this case as follows.

Theorem 13. For any analytic sheaf homomorphism $\varphi: {}_V\mathcal{O}^r \longrightarrow {}_V\mathcal{O}^s$ over an analytic variety V, the kernel of φ is a finitely generated analytic sheaf in an open neighborhood of any point of V.

Proof. Since the desired result is local in character, it suffices to consider an analytic subvariety V of an open neighborhood U of the origin in $\mathbf{C}^n$, representing a neighborhood of some fixed point on the given variety. Letting $\mathcal{J}(V) \subset {}_n\mathcal{O}|U$ be the sheaf of ideals of the subvariety $V \subset U$, both $\mathcal{J}(V)$ and the residue class sheaf ${}_V\tilde{\mathcal{O}} = ({}_n\mathcal{O}|U)/\mathcal{J}(V)$ are coherent analytic sheaves over the open set U. Note that the restriction ${}_V\tilde{\mathcal{O}}|V$ is precisely the sheaf ${}_V\mathcal{O}$, considered merely as a sheaf of rings. The sheaf homomorphism $\varphi: {}_V\mathcal{O}^r \longrightarrow {}_V\mathcal{O}^s$ can as usual be represented by a matrix whose coefficients m_{ij} are holomorphic functions on the subvariety V; and if the neighborhood U is chosen sufficiently small, there will exist holomorphic functions M_{ij} in U such that $M_{ij}|V = m_{ij}$. The matrix (M_{ij}) determines a homomorphism of analytic sheaves $\Phi: ({}_n\mathcal{O}|U)^r \longrightarrow ({}_n\mathcal{O}|U)^s$; and since this homomorphism evidently takes the submodule $\mathcal{J}(V)^r \subset ({}_n\mathcal{O}|U)^r$ into the submodule $\mathcal{J}(V)^s \subset ({}_n\mathcal{O}|U)^s$, it induces on the quotient sheaves a homomorphism $\tilde{\varphi}: {}_V\tilde{\mathcal{O}}^r \longrightarrow {}_V\tilde{\mathcal{O}}^s$.

Note that the restriction $\tilde{\Phi}|V$ is precisely the original homomorphism φ. Now since ${}_V\tilde{\mathcal{O}}$ is a coherent analytic sheaf in U, the kernel of the homomorphism $\tilde{\Phi}$ is also a coherent analytic sheaf in U; and therefore, after shrinking the neighborhood U if necessary, there will exist a sheaf homomorphism Ψ such that the following is an exact sequence of analytic sheaves in U :

$$({}_n\mathcal{O}|U)^q \xrightarrow{\Psi} {}_V\tilde{\mathcal{O}}^r \xrightarrow{\tilde{\Phi}} {}_V\tilde{\mathcal{O}}^s .$$

Note that the homomorphism Ψ can be factored into a product $\Psi = \tilde{\Psi}\Psi_o$ where Ψ_o is the natural mapping $({}_n\mathcal{O}|U)^q \rightarrow {}_V\tilde{\mathcal{O}}^q$ and $\tilde{\Psi}$: ${}_V\tilde{\mathcal{O}}^q \rightarrow {}_V\tilde{\mathcal{O}}^r$; and since Ψ_o is surjective there results the further exact sequence of analytic sheaves in U :

$${}_V\tilde{\mathcal{O}}^q \xrightarrow{\tilde{\Psi}} {}_V\tilde{\mathcal{O}}^r \xrightarrow{\tilde{\Phi}} {}_V\tilde{\mathcal{O}}^s .$$

Restricting the latter exact sequence to V yields an exact sequence of analytic sheaves over the variety V of the form

$${}_V\mathcal{O}^q \xrightarrow{\Psi} {}_V\mathcal{O}^r \xrightarrow{\varphi} {}_V\mathcal{O}^s ,$$

so that the kernel of φ is locally finitely generated over ${}_V\mathcal{O}$ and the proof is thereby concluded.

An analytic sheaf $\mathcal{S}$ over an analytic variety V is said to be <u>coherent</u> if in some open neighborhood W_a of each point $a \in V$ there is an exact sequence of analytic sheaves over V of the form

$$({}_V\mathcal{O}|W_a)^r \longrightarrow ({}_V\mathcal{O}|W_a)^s \longrightarrow (\mathcal{S}|W_a) \longrightarrow 0 ,$$

for some r, s . It then follows from Oka's theorem, as in the case of sheaves over open subsets of $\mathbb{C}^n$, that coherence is preserved under the usual algebraic operations on sheaves.

When the variety V is represented as an analytic subvariety of an open set $U \subseteq \mathbb{C}^n$, there have then been introduced two separate notions of coherence, one for sheaves of modules over the sheaf of rings ${}_n\mathcal{O}$ in U and another for sheaves of modules over the sheaf of rings ${}_V\mathcal{O}$ over V ; and perhaps a few words of comparison are in order here. On the one hand, if $\mathcal{S}$ is an analytic sheaf in the open set $U \subseteq \mathbb{C}^n$ and if the ideal $\mathcal{J}(V)_a \subset {}_n\mathcal{O}_a$ acts trivially on the module $\mathcal{S}_a$ for each point $a \in V$, then of course the restriction $\mathcal{S}|V$ can be viewed as a sheaf of modules over the sheaf of rings ${}_V\mathcal{O}$, that is, as an analytic sheaf on the variety V . If in addition $\mathcal{S}$ is a coherent analytic sheaf in U , then it is easy to see that the restriction $\mathcal{S}|V$ is a coherent analytic sheaf on the variety V . For suppose there is an exact sequence of sheaves of ${}_n\mathcal{O}$-modules over U of the form

$$({}_n\mathcal{O}|U)^r \xrightarrow{\Psi} ({}_n\mathcal{O}|U)^s \xrightarrow{\Phi} \mathcal{S} \longrightarrow 0 .$$

Considering any point $a \in V$, since by assumption the ideal $\mathcal{J}(V)_a$ acts trivially on $\mathcal{S}_a$, it follows readily that $\Phi(\mathcal{J}(V)_a^s) = 0 \in \mathcal{S}_a$ and hence that Φ induces a homomorphism $\varphi\colon {}_V\mathcal{O}_a^s \to \mathcal{S}_a$ with image the full module $\mathcal{S}_a$. If $\varphi(f) = 0$

for some element $f \in {}_V\mathcal{O}_a^s$, then $\Phi(F) = 0$ for any element $F \in {}_n\mathcal{O}_a^s$, representing the residue class f , and so necessarily $F = \Psi(G)$ for some element $G \in {}_n\mathcal{O}_a^r$; letting $\psi\colon {}_V\mathcal{O}^r \longrightarrow {}_V\mathcal{O}^s$ be the mapping naturally induced by Ψ , it follows that $f = \psi(g)$ where $g \in {}_V\mathcal{O}_a^r$ is the residue class of $G \in {}_n\mathcal{O}_a^r$. Thus there results an exact sequence of analytic sheaves over V of the form

$${}_V\mathcal{O}^r \xrightarrow{\ \psi\ } {}_V\mathcal{O}^s \xrightarrow{\ \varphi\ } \mathcal{S}|V \longrightarrow 0 ,$$

so $\mathcal{S}|V$ is a coherent analytic sheaf over V , as desired. On the other hand, if $\mathcal{S}$ is an analytic sheaf over the variety V , it can of course be viewed as a sheaf of modules over the sheaf of rings ${}_n\mathcal{O}|V$. Introduce over the open set U the sheaf $\tilde{\mathcal{S}}$ with stalks defined by

$$\tilde{\mathcal{S}}_a = \begin{cases} \mathcal{S}_a , \text{ viewed as an } {}_n\mathcal{O}_a\text{-module, for } a \in V \\ 0 , \text{ for } a \in U-V , \end{cases}$$

and with topology defined by taking as sections over an open subset $W \subseteq U$ the sections of $\mathcal{S}$ over $W \cap V$ extended by the zero section over $W - W \cap V$. This is called the <u>trivial extension</u> of the sheaf $\mathcal{S}$ to the open set $U \subseteq \mathbb{C}^n$, and is readily seen to be an analytic sheaf over U . In particular, for the sheaf ${}_V\mathcal{O}$ itself the trivial extension is clearly the coherent analytic sheaf ${}_V\tilde{\mathcal{O}} = ({}_n\mathcal{O}|U)/\mathcal{J}(V)$ over the open set U . Note that for any homomorphism $\varphi\colon \mathcal{R} \longrightarrow \mathcal{S}$ between analytic sheaves over the subvariety V , there is an obvious induced homomorphism $\Phi\colon \tilde{\mathcal{R}} \longrightarrow \tilde{\mathcal{S}}$

between the trivial extensions of these sheaves over the open set U ; and that whenever

$$\mathcal{R} \xrightarrow{\varphi} \mathcal{S} \xrightarrow{\psi} \mathcal{T}$$

is an exact sequence of analytic sheaves over the subvariety V , the induced homomorphisms between the trivial extensions of these various sheaves form the exact sequence

$$\tilde{\mathcal{R}} \xrightarrow{\Phi} \tilde{\mathcal{S}} \xrightarrow{\Psi} \tilde{\mathcal{T}}$$

of analytic sheaves over the open set U . It is an immediate consequence of these remarks that for any coherent analytic sheaf $\mathcal{S}$ over the subvariety V , the trivial extension $\tilde{\mathcal{S}}$ is a coherent analytic sheaf over the open set U ; indeed, since $\tilde{\mathcal{S}} | V = \mathcal{S}$, the sheaf $\mathcal{S}$ is coherent over V if and only if its trivial extension $\tilde{\mathcal{S}}$ is coherent over U .

(d) Suppose that V_1 and V_2 are irreducible germs of analytic subvarieties which determine equivalent germs of analytic varieties; it is then clear that considering analytic subvarieties V_1, V_2 representing these germs, the regular loci $\mathcal{R}(V_1)$ and $\mathcal{R}(V_2)$ are complex analytic manifolds of the same dimension, and consequently that $\dim V_1 = \dim V_2$. Thus it is possible to define the <u>dimension</u> of an irreducible germ of an analytic variety to be the dimension of any representative analytic subvariety. Having done so, the obvious elementary properties of dimension carry over to analytic varieties almost immediately. For an arbitrary germ V

of an analytic variety, written as the union of its irreducible components $V = \cup_i V_i$, the dimension of the germ V will be defined by $\dim V = \max_i \dim V_i$; so again the dimension of V is just the dimension of any representative subvariety. The germ V will be called pure dimensional if $\dim V = \dim V_i$ for all the irreducible components V_i of V . The dimension of an analytic variety at a point will be defined to be the dimension of the germ of the variety at that point; and the dimension of the analytic variety as a whole will be defined to be the maximum dimension of the variety at all of its points. An analytic variety will be said to be pure dimensional if its germs at all points are pure dimensional and of a constant dimension, which must of course be the dimension of the analytic variety as a whole. Note that an analytic subvariety of any analytic variety can be viewed as an analytic variety itself, so has a well defined dimension. For an analytic variety V of pure dimension k , it is clear that the regular locus $\mathcal{R}(V)$ is a k-dimensional complex analytic manifold, and is a dense open subset of V , although not necessarily a connected subset; and that the singular locus $\mathcal{S}(V)$ is an analytic subvariety with $\dim \mathcal{S}(V) < k$, although not necessarily a pure dimensional subvariety.

Recalling the results of §3(d), note that if V is an irreducible germ of an analytic variety and $W \subset V$ is the germ of an analytic subvariety of V , then necessarily $\dim W < \dim V$. Furthermore, if V is the germ of an analytic variety of pure dimension k and $f \in {}_V\mathcal{O}$ is a non-unit which does not vanish identically on any component of V , then the germ of the subvariety $\{z \in V | f(z) = 0\}$ is of pure dimension k-1 . However, as will be seen later, the converse assertion does not necessarily hold; a subvariety of V of pure dimension k-1 cannot necessarily be described as the locus of zeros of a single function in ${}_V\mathcal{O}$.

It is clearly of some interest to characterize the dimension of the germ V of an analytic variety directly in terms of the local ring ${}_V\mathcal{O}$; it is indeed possible to do so quite simply, by means of the following purely algebraic concepts. For an arbitrary ring $\mathcal{O}$, the depth of a prime ideal $\mathfrak{p} \subset \mathcal{O}$ is defined to be the largest integer d such that there exist prime ideals $\mathfrak{p}_i \subset \mathcal{O}$ for which

$$\mathfrak{p} = \mathfrak{p}_o \subset \mathfrak{p}_1 \subset \mathfrak{p}_2 \subset \cdots \subset \mathfrak{p}_d ;$$

the depth of the ideal $\mathfrak{p}$ will be denoted by depth $\mathfrak{p}$, and is a non-negative integer or ∞ . In a complementary fashion, the height of a prime ideal $\mathfrak{p} \subset \mathcal{O}$ is defined to be the largest integer h such that there exist prime ideals $\mathfrak{p}_i \subset \mathcal{O}$ for which

$$\mathfrak{p} = \mathfrak{p}_o \supset \mathfrak{p}_1 \supset \mathfrak{p}_2 \supset \cdots \supset \mathfrak{p}_h ;$$

the height of the ideal $\mathfrak{p}$ will be denoted by height $\mathfrak{p}$, and is also a non-negative integer or ∞ . It should be emphasized that all these containments are proper containments, and that all the prime ideals considered are properly contained in the ring $\mathcal{O}$. Note that a prime ideal has depth 0 if and only if it is a maximal ideal; and note further that when the ring $\mathcal{O}$ is an integral domain, the zero ideal is a prime ideal, and is the unique prime ideal of height 0 .

Theorem 14(a). For a prime ideal $\mathcal{Y} \subset {}_n\mathcal{O}$ determining an irreducible germ $V = \text{loc}\,\mathcal{Y}$ of an analytic subvariety at the origin in $\mathbb{C}^n$, it follows that

$$\text{depth}\,\mathcal{Y} = \dim V$$

$$\text{height}\,\mathcal{Y} = n - \dim V .$$

Remark. The difference $n-\dim V$ appears quite frequently, and as a convenient abbreviation it will be called the codimension of the germ of subvariety $V \subset \mathbb{C}^n$ and will be denoted by codim V ; thus the last conclusion of the theorem is that height $\mathcal{Y}$ = codim V .

Proof. As a preliminary observation, note that if $W' \subset W$ are any irreducible germs of analytic subvarieties at the origin in $\mathbb{C}^n$ such that $\dim W' < \dim W-1$, then there exists an irreducible germ W'' of analytic subvariety such that $W' \subset W'' \subset W$. To see this, select any element $f \in \text{id}\, W' - \text{id}\, W \subset {}_n\mathcal{O}$, noting that $\text{id}\, W \subset \text{id}\, W'$; and consider the germ of analytic subvariety $W_o = \{z \in W | f(z) = 0\}$, noting that $W' \subseteq W_o \subset W$. It is clear that for some irreducible component W'' of the germ W_o , necessarily $W' \subseteq W'' \subset W$; but it follows from Theorem 9(e) that $\dim W'' = \dim W-1 > \dim W'$, and hence that $W' \subset W'' \subset W$.

Turning now to the proof of the theorem itself, let $k = \dim V$, $d = \text{depth}\,\mathcal{Y}$, and $h = \text{height}\,\mathcal{Y}$. It follows immediately from the observation made in the preceding paragraph that there are irreducible germs V_i of analytic subvarieties at the origin in $\mathbb{C}^n$ such that $\dim V_i = i$ and that

$$V_0 \subset V_1 \subset \cdots \subset V_{k-1} \subset V \subset V_{k+1} \subset V_{k+2} \subset \cdots \subset V_{n-1} \subset V_n .$$

The ideals $\mathfrak{P}_i = \text{id}\, V_i \subset {}_n\mathcal{O}$ are then proper prime ideals for which

$$\mathfrak{P}_0 \supset \mathfrak{P}_1 \supset \cdots \supset \mathfrak{P}_{k-1} \supset \mathfrak{P} \supset \mathfrak{P}_{k+1} \supset \mathfrak{P}_{k+2} \supset \cdots \supset \mathfrak{P}_{n-1} \supset \mathfrak{P}_n ;$$

and consequently it is clear that

$$\text{depth}\, \mathfrak{P} \geqq k \quad \text{and} \quad \text{height}\, \mathfrak{P} \geqq n-k .$$

On the other hand, there must exist some prime ideals $\mathfrak{P}'_i \subset {}_n\mathcal{O}$ and $\mathfrak{P}''_i \subset {}_n\mathcal{O}$ such that

$$\mathfrak{P}'_h \subset \cdots \subset \mathfrak{P}'_1 \subset \mathfrak{P} \subset \mathfrak{P}''_1 \subset \cdots \subset \mathfrak{P}''_d ;$$

and the irreducible germs $V'_i = \text{loc}\, \mathfrak{P}'_i$ and $V''_i = \text{loc}\, \mathfrak{P}''_i$ of analytic subvarieties at the origin in $\mathbb{C}^n$ are then such that

$$V'_h \supset \cdots \supset V'_1 \supset V \supset V''_1 \supset \cdots \supset V''_d .$$

Since the dimensions of any two consecutive subvariety germs in this chain differ by at least 1, as a consequence of Theorem 9(a), it follows readily that

$$n \geqq \dim V'_h \geqq k+h \quad \text{and} \quad 0 \leqq \dim V''_d \leqq \dim V - d ,$$

or equivalently that

$$\text{height}\, \mathfrak{P} \leqq n-k \quad \text{and} \quad \text{depth}\, \mathfrak{P} \leqq k .$$

Comparing these two sets of inequalities yields the desired result.

It is of course a trivial consequence of this theorem that depth $\mathfrak{p}$ + height $\mathfrak{p}$ = n for any prime ideal $\mathfrak{p} \subset {}_n\mathcal{O}$. Several other consequences follow almost immediately from previously established properties of dimension, and should perhaps be mentioned in passing. The locus of the zero ideal is all of $\mathbb{C}^n$, hence is an n-dimensional analytic subvariety; so that depth 0 = n and height 0 = 0. The locus of the maximal ideal ${}_n\mathfrak{m} \subset {}_n\mathcal{O}$ is just the origin in $\mathbb{C}^n$, a 0-dimensional analytic subvariety; so that depth ${}_n\mathfrak{m}$ = 0 and height ${}_n\mathfrak{m}$ = n. Theorem 9(c) asserts that a prime ideal $\mathfrak{p} \subset {}_n\mathcal{O}$ is a principal ideal precisely when dim loc $\mathfrak{p}$ = n-1; so that <u>a prime ideal</u> $\mathfrak{p} \subset {}_n\mathcal{O}$ <u>is principal if and only if</u> height $\mathfrak{p}$ = 1, or equivalently if and only if depth $\mathfrak{p}$ = n-1. It follows from Theorem 9(f) that when the prime ideal $\mathfrak{p}$ is generated by r elements of ${}_n\mathcal{O}$ then necessarily dim loc $\mathfrak{p} \geqq$ n-r; so that <u>if a prime ideal</u> $\mathfrak{p} \subset {}_n\mathcal{O}$ <u>is generated by</u> r <u>elements then</u> height $\mathfrak{p} \leqq$ r, or equivalently depth $\mathfrak{p} \geqq$ n-r.

<u>Theorem 14(b).</u> For an irreducible germ V of an analytic variety

$$\text{height } {}_V\mathfrak{m} = \dim V$$

where ${}_V\mathfrak{m}$ is the maximal ideal of the local ring ${}_V\mathcal{O}$.

Proof. Represent the germ V of an analytic variety by the germ V of an analytic subvariety at the origin in $\mathbb{C}^n$, and let $\mathfrak{p}$ = id V $\subset {}_n\mathcal{O}$; thus $\mathfrak{p}$ is a prime ideal in ${}_n\mathcal{O}$, and ${}_V\mathcal{O} = {}_n\mathcal{O}/\mathfrak{p}$. Note that for any prime ideal $\mathfrak{p}_i \subset {}_n\mathcal{O}$ for which $\mathfrak{p} \subseteq \mathfrak{p}_i$, the residue class ideal $\tilde{\mathfrak{p}}_i = \mathfrak{p}_i/\mathfrak{p} \subset {}_V\mathfrak{O}$ is a prime ideal in ${}_V\mathfrak{O}$; it is evident that there thus arises a one-to-one correspondence between the set of those prime ideals in ${}_n\mathfrak{O}$ containing $\mathfrak{p}$ and the set of all prime ideals in ${}_V\mathfrak{O}$. If d is the depth of the prime ideal $\mathfrak{p}$ in ${}_n\mathfrak{O}$, there are prime ideals $\mathfrak{p}_i \subset {}_n\mathfrak{O}$ such that $\mathfrak{p} = \mathfrak{p}_o \subset \mathfrak{p}_1 \subset \cdots \subset \mathfrak{p}_d$; and it is evident that $\mathfrak{p}_d = {}_n\mathfrak{m}$, the maximal ideal of the local ring ${}_n\mathfrak{O}$. Passing to the residue class ring ${}_V\mathfrak{O}$, and noting that ${}_V\mathfrak{m} = {}_n\tilde{\mathfrak{m}} = \tilde{\mathfrak{p}}_d$, there results a chain of prime ideals in ${}_V\mathfrak{O}$ of the form $\tilde{\mathfrak{p}}_o \subset \tilde{\mathfrak{p}}_1 \subset \cdots \subset \tilde{\mathfrak{p}}_d = {}_V\mathfrak{m}$; and consequently $d \leqq \text{height } {}_V\mathfrak{m}$. On the other hand, if h is the height of the maximal ideal ${}_V\mathfrak{m}$ in ${}_V\mathfrak{O}$, there are prime ideals $\tilde{\mathfrak{p}}'_i \subset {}_V\mathfrak{O}$ such that ${}_V\mathfrak{m} = \tilde{\mathfrak{p}}'_o \supset \tilde{\mathfrak{p}}'_1 \supset \cdots \supset \tilde{\mathfrak{p}}'_h$, and since ${}_V\mathfrak{O}$ is an integral domain it is evident that $\tilde{\mathfrak{p}}'_h = 0$, the zero ideal of ${}_V\mathfrak{O}$. Writing $\tilde{\mathfrak{p}}'_i = \mathfrak{p}_i/\mathfrak{p}$ for some prime ideals $\mathfrak{p} \subseteq \mathfrak{p}_i \subset {}_n\mathfrak{O}$, and noting that $\tilde{\mathfrak{p}}'_h = \mathfrak{p}/\mathfrak{p}$, there results a chain of prime ideals in ${}_n\mathfrak{O}$ of the form $\mathfrak{p}'_o \supset \mathfrak{p}'_1 \supset \cdots \supset \mathfrak{p}'_h = \mathfrak{p}$; and consequently $h \leqq \text{depth } \mathfrak{p}$. The two inequalities just established show that $\text{height } {}_V\mathfrak{m} = \text{depth } \mathfrak{p}$; and since $\dim V = \text{depth } \mathfrak{p}$ by Theorem 14(a), the desired result has been established.

The <u>Krull dimension</u> of a ring $\mathfrak{O}$ is defined to be the largest integer k for which there exist $k+1$ prime ideals $\mathfrak{p}_i \subset \mathfrak{O}$ such that

$$\mathfrak{p}_0 \subset \mathfrak{p}_1 \subset \mathfrak{p}_2 \subset \cdots \subset \mathfrak{p}_k ;$$

and it is denoted by Krull dim $\mathcal{O}$. For a local ring $\mathcal{O}$ with maximal ideal $\mathfrak{m}$, it is clear that Krull dim $\mathcal{O}$ = height $\mathfrak{m}$; for the longest chain of prime ideals of the above form evidently must end with the maximal ideal, $\mathfrak{p}_k = \mathfrak{m}$. For an integral domain $\mathcal{O}$ the zero ideal is a prime ideal, and it is clear that Krull dim $\mathcal{O}$ = depth 0 ; for the longest chain of prime ideals of the above form evidently must begin with the zero ideal, $\mathfrak{p}_0 = 0$. For the local integral domain ${}_V\mathcal{O}$ of an irreducible germ V of an analytic variety, then, Krull dim ${}_V\mathcal{O}$ = height ${}_V\mathfrak{m}$ = depth 0 ; and the last theorem can be restated as the equality

$$\dim V = \text{Krull dim } {}_V\mathcal{O}$$

for any irreducible germ V of an analytic variety.

(e) Any germ V of an analytic variety can be represented by a germ of an analytic subvariety at the origin in the complex number space $\mathbb{C}^n$ of some dimension n; the smallest dimension n for which such a representation is possible will be called the <u>imbedding dimension</u> of the analytic variety V , and will be denoted by imbed dim V . It is of course clear that imbed dim V $\geqq$ dim V for any germ V of a complex analytic variety; and it is also clear that imbed dim V = = dim V if and only if V is the germ of a regular analytic variety, a complex analytic manifold. The imbedding dimension can be considerably larger than the dimension for some varieties, though;

indeed, the imbedding dimension cannot be bounded by any function of the dimension, even for irreducible analytic varieties. It is of evident interest to characterize the imbedding dimension of a germ of an analytic variety V in terms of other properties of V, in particular in terms of the local ring ${}_V\mathcal{O}$ of the variety. Such a characterization is quite easy, in purely algebraic terms; but it is convenient first to establish the following useful auxiliary result.

Nakayama's Lemma. Let $\mathcal{O}$ be an arbitrary local ring with maximal ideal $\mathfrak{m}$. If M and $N \subseteq M$ are modules over the ring $\mathcal{O}$ such that the quotient module M/N is finitely generated and $M = N + \mathfrak{m}\cdot M$, then necessarily $N = M$.

Proof. Passing to the quotient module $L = M/N$, it clearly suffices to show that if L is a finitely generated module over the ring $\mathcal{O}$ such that $L = \mathfrak{m}\cdot L$, then necessarily $L = 0$. If $L \neq 0$, choose a minimal set of generators $X_1,\ldots,X_r$ of the module L; thus the elements $X_1,\ldots,X_r$ generate L, but no proper subset of these elements serve to generate L. Since $L = \mathfrak{m}\cdot L$, there must exist elements $m_i \in \mathfrak{m}$ such that $X_r = \sum_{i=1}^{r} m_i X_i$, hence such that $(1-m_r)\cdot X_r = \sum_{i=1}^{r-1} m_i X_i$; but if $m_r \in \mathfrak{m}$ then $(1-m_r) \notin \mathfrak{m}$, so that $(1-m_r)$ is invertible in the ring $\mathcal{O}$ and $X_r = \sum_{i=1}^{r-1} m_i(1-m_r)^{-1}\cdot X_i$. That then shows that $X_1,\ldots,X_{r-1}$ serve to generate the module L; and with this contradiction, the proof is concluded.

Remark. Properly speaking, Nakayama's Lemma is rather more general than the result just established; but since the more general result will not be required at present, the preceding misattribution will be used for terminological convenience. Note that as a special case of the lemma, if $\mathfrak{A}$ and $\mathfrak{B} \subseteq \mathfrak{A}$ are two ideals in the ring $\mathcal{O}$ such that $\mathfrak{A} = \mathfrak{B} + \mathfrak{m} \cdot \mathfrak{A}$, then necessarily $\mathfrak{B} = \mathfrak{A}$.

Theorem 15. The imbedding dimension of a germ V of an analytic variety is the minimal number of generators of the maximal ideal ${}_V\mathfrak{m}$ of the local ring ${}_V\mathcal{O}$.

Proof. Select a germ V of an analytic subvariety at the origin in $\mathbb{C}^n$, representing the given germ of an analytic variety. On the one hand, note that the coordinate functions $z_1, \ldots, z_n$ in $\mathbb{C}^n$ generate the maximal ideal ${}_n\mathfrak{m} \subset {}_n\mathcal{O}$, hence that the residue classes $\tilde{z}_1, \ldots, \tilde{z}_n$ in ${}_V\mathcal{O}$ generate the maximal ideal ${}_V\mathfrak{m} \subset {}_V\mathcal{O}$; it is consequently obvious that the minimal number of generators of the maximal ideal ${}_V\mathfrak{m} \subset {}_V\mathcal{O}$ is less than or equal to the imbedding dimension of V. On the other hand, suppose that $f_1, \ldots, f_m$ are analytic functions at the origin in $\mathbb{C}^n$ such that the residue classes $\tilde{f}_1, \ldots, \tilde{f}_m$ in ${}_V\mathcal{O}$ are a set of generators of the maximal ideal ${}_V\mathfrak{m} \subset {}_V\mathcal{O}$, and that m is the minimal number of generators of the ideal ${}_V\mathfrak{m}$. Note that the differentials $df_1(0), \ldots, df_m(0)$ are necessarily linearly independent vectors in $\mathbb{C}^n$. (Otherwise, after relabeling these functions if necessary, there exist complex constants c_i such that $f_1 = c_2 f_2 + \ldots + c_m f_m + r$

for some function $r \in {}_n\mathcal{O}$ with $dr(0) = 0$, or equivalently, with $r \in {}_n\mathcal{M}^2$. Letting $\mathcal{N} \subset {}_V\mathcal{O}$ be the ideal generated by the residue classes $\tilde{f}_2,\ldots,\tilde{f}_m$ in ${}_V\mathcal{O}$, and recalling that $\tilde{f}_1,\tilde{f}_2,\ldots,\tilde{f}_m$ generate the ideal ${}_V\mathcal{M} \subset {}_V\mathcal{O}$, it follows from the above equality that ${}_V\mathcal{M} = \mathcal{N} + {}_V\mathcal{O}\cdot\tilde{f}_1 = \mathcal{N} + {}_V\mathcal{O}\cdot\tilde{r} \subseteq \mathcal{N} + {}_V\mathcal{M}^2 \subseteq {}_V\mathcal{M}$; hence from Nakayama's Lemma it follows that $\mathcal{N} = {}_V\mathcal{M}$, so that actually $\tilde{f}_2,\ldots,\tilde{f}_m$ generate the ideal ${}_V\mathcal{M}$. This contradicts the hypothesis that m is the minimal number of generators of the ideal ${}_V\mathcal{M} \subset {}_V\mathcal{O}$ hence serves to complete the proof of the assertion.) The functions $f_1,\ldots,f_m$ can then be taken as part of a set of coordinate functions in $\mathbb{C}^n$; so there is no loss of generality in setting $z_1 = f_1,\ldots,$ $z_m = f_m$. Now for any other coordinate function z_i , $i = m+1,\ldots,n$, the residue class $\tilde{z}_i$ is contained in the maximal ideal ${}_V\mathcal{M}$, hence $\tilde{z}_i = \tilde{g}_{i1}\tilde{z}_1 + \ldots \tilde{g}_{im}\tilde{z}_m$ for some analytic functions $g_{ij} \in {}_n\mathcal{O}$, since $\tilde{z}_1 = \tilde{f}_1,\ldots,\tilde{z}_m = \tilde{f}_m$ generate ${}_V\mathcal{M}$. The functions

$$w_1 = z_1,\ldots, \quad w_m = z_m, \quad w_i = z_i - g_{i1}z_1 - \cdots - g_{im}z_m \qquad (i = m+1,\ldots,n)$$

are also coordinate functions in a neighborhood of the origin in $\mathbb{C}^n$; and by construction, the residue class $\tilde{w}_i = 0$ in ${}_V\mathcal{O}$, or equivalently, the coordinate function w_i vanishes identically on V , for $i = m+1,\ldots,n$. Thus $V \subseteq \{(w_1,\ldots,w_n) \in \mathbb{C}^n \mid w_{m+1} = \cdots = w_n = 0\}$, so that imbed dim $V \leqq m$. The theorem is thereby proved.

The proof of the preceding theorem suggests various other expressions for the imbedding dimension of a germ V of an analytic variety. First, in terms of the maximal ideal ${}_V\mathcal{M}$ itself, consider the ideals ${}_V\mathfrak{m}^2 \subset {}_V\mathfrak{m}$ as modules over the ring ${}_V\mathcal{O}$, and introduce the quotient module ${}_V\mathfrak{m}/{}_V\mathfrak{m}^2$. Since clearly $f\cdot({}_V\mathfrak{m}/{}_V\mathfrak{m}^2) = 0$ for any $f \in {}_V\mathfrak{m}$, it follows that ${}_V\mathfrak{m}/{}_V\mathfrak{m}^2$ can be viewed as a module over the residue class ring ${}_V\mathcal{O}/{}_V\mathfrak{m}$; but since ${}_V\mathcal{O}/{}_V\mathfrak{m}$ is just the complex number field, this is equivalent to viewing ${}_V\mathfrak{m}/{}_V\mathfrak{m}^2$ as a complex vector space. The dimension $\dim_{\mathbf{C}}\, {}_V\mathfrak{m}/{}_V\mathfrak{m}^2$ of this vector space is of course finite, since the original ideal ${}_V\mathfrak{m}$ is necessarily finitely generated.

Corollary 1 to Theorem 15. For any germ V of an analytic variety,

$$\text{imbed dim } V = \dim_{\mathbf{C}}\, {}_V\mathfrak{m}/{}_V\mathfrak{m}^2 .$$

Proof. If imbed dim $V = n$ there are n functions $f_1,\ldots,f_n \in {}_V\mathfrak{m}$ which generate the maximal ideal ${}_V\mathfrak{m}$, as a consequence of Theorem 15; the images of these functions in the quotient module ${}_V\mathfrak{m}/{}_V\mathfrak{m}^2$ then generate that module as well, so that $\dim_{\mathbf{C}}\, {}_V\mathfrak{m}/{}_V\mathfrak{m}^2 \leqq n = \text{imbed dim } V$. On the other hand if $\dim_{\mathbf{C}}\, {}_V\mathfrak{m}/{}_V\mathfrak{m}^2 = m$ there are m functions $g_1,\ldots,g_m \in {}_V\mathfrak{m}$ such that the images of these functions in the quotient module ${}_V\mathfrak{m}/{}_V\mathfrak{m}^2$ generate that module. These functions $g_1,\ldots,g_m$ consequently generate an ideal $\mathfrak{n} \subseteq {}_V\mathfrak{m}$, which evidently has the property that $\mathfrak{n} + {}_V\mathfrak{m}^2 = {}_V\mathfrak{m}$; but from Nakayama's Lemma it follows that $\mathfrak{n} = {}_V\mathfrak{m}$, hence that ${}_V\mathfrak{m}$ has m generators. Referring to Theorem 15 again, necessarily imbed dim $V \leqq m = \dim_{\mathbf{C}}\, {}_V\mathfrak{m}/{}_V\mathfrak{m}^2$; and that suffices to conclude the proof.

The imbedding dimension of an analytic variety can also be expressed in terms of the ideal of a representative analytic subvariety.

Corollary 2 to Theorem 15. Let V be the germ of an analytic subvariety at the origin in $\mathbf{C}^n$, and let $f_1,\ldots,f_m \in {}_n\mathcal{O}$ be generators for the ideal $\mathrm{id}\, V \subset {}_n\mathcal{O}$. If the subspace of $\mathbf{C}^n$ spanned by the vectors $df_1(0),\ldots,df_m(0)$ is of dimension r, then for the germ V of an analytic variety represented by the given germ of a subvariety it follows that

$$\text{imbed dim } V = n - r .$$

Proof. Viewing the ideal $\mathfrak{A} = \mathrm{id}\, V \subset {}_n\mathcal{O}$ as well as the maximal ideals ${}_n\mathfrak{m} \subset {}_n\mathcal{O}$ and ${}_V\mathfrak{m} \subset {}_V\mathcal{O}$ as modules over the ring ${}_n\mathcal{O}$, the following is an exact sequence of ${}_n\mathcal{O}$-modules:

$$0 \longrightarrow \mathfrak{A} \longrightarrow {}_n\mathfrak{m} \longrightarrow {}_V\mathfrak{m} \longrightarrow 0 .$$

The submodule ${}_n\mathfrak{m}^2 \subset {}_n\mathfrak{m}$ is mapped onto the submodule ${}_V\mathfrak{m}^2 \subset {}_V\mathfrak{m}$, hence there results the following exact sequence of quotient modules over the ring ${}_n\mathcal{O}$:

$$0 \longrightarrow \frac{\mathfrak{A} + {}_n\mathfrak{m}^2}{{}_n\mathfrak{m}^2} \longrightarrow \frac{{}_n\mathfrak{m}}{{}_n\mathfrak{m}^2} \longrightarrow \frac{{}_V\mathfrak{m}}{{}_V\mathfrak{m}^2} \longrightarrow 0 .$$

It is clear that any element $f \in {}_n\mathfrak{m} \subset {}_n\mathcal{O}$ acts as the zero element on each of these ${}_n\mathcal{O}$-modules, so that this last sequence can indeed be viewed as an exact sequence of modules over the residue class ring ${}_n\mathcal{O}/{}_n\mathfrak{m}$, hence as an exact sequence of complex vector spaces since ${}_n\mathcal{O}/{}_n\mathfrak{m} \cong \mathbb{C}$. The vector space structure of ${}_V\mathfrak{m}/{}_V\mathfrak{m}^2$ is of course the same as that obtained by viewing ${}_V\mathfrak{m}/{}_V\mathfrak{m}^2$ as a module over the ring ${}_V\mathcal{O}$, so from Corollary 1 it follows that $\dim_{\mathbb{C}} {}_V\mathfrak{m}/{}_V\mathfrak{m}^2$ = imbed dim V ; and since $\dim_{\mathbb{C}} {}_n\mathfrak{m}/{}_n\mathfrak{m}^2 = n$, the exact sequence of vector spaces leads to the identity

$$\text{imbed dim } V = n - \dim_{\mathbb{C}} \frac{\mathfrak{A} + {}_n\mathfrak{m}^2}{{}_n\mathfrak{m}^2} .$$

Now the mapping which associates to any function $f \in {}_n\mathfrak{m}$ its differential $df(0) \in \mathbb{C}^n$ is obviously a linear mapping with kernel precisely ${}_n\mathfrak{m}^2$; hence this mapping can be used to identify ${}_n\mathfrak{m}/{}_n\mathfrak{m}^2$ with the complex vector space $\mathbb{C}^n$. The subspace $(\mathfrak{A} + {}_n\mathfrak{m}^2)/{}_n\mathfrak{m}^2 \subseteq {}_n\mathfrak{m}/{}_n\mathfrak{m}^2$ is generated by the functions $f_1,\ldots,f_m \in {}_n\mathcal{O}$; so under this identification it becomes the subspace of $\mathbb{C}^n$ spanned by the vectors $df_1(0),\ldots,df_m(0)$, hence is a complex vector space of dimension r . Consequently imbed dim V = n-r , as asserted.

A germ V of an analytic subvariety at the origin in $\mathbb{C}^n$ will be called a <u>neat</u> germ of a subvariety if imbed dim V = n for the germ V of an analytic variety represented by that germ of subvariety; occasionally, as a convenient if not wholly accurate terminology, a neat germ of a subvariety will be called a <u>neat imbedding</u> of the germ of an analytic subvariety it represents. A neat imbedding is neat in the sense that it represents the given analytic variety as a subvariety of the ambient complex number space of least possible dimension, hence with no waste or excess. It follows immediately from Corollary 2 to Theorem 15 that neat germs of subvarieties can be characterized in the following manner.

Corollary 3 to Theorem 15. A germ V of an analytic subvariety at the origin in $\mathbb{C}^n$ is neat if and only if $df(0) = 0$ for all elements $f \in \text{id } V \subset {}_n\mathcal{O}$.

Neat germs of subvarieties form a very convenient class of subvarieties for a number of reasons, such as the following.

Corollary 4 to Theorem 15. Two neat germs V, V' of analytic subvarieties at the origin in $\mathbb{C}^n$ determine equivalent germs of analytic varieties if and only if they are equivalent germs of analytic subvarieties.

Proof. Of course if V and V' determine equivalent germs of analytic subvarieties they determine equivalent germs of analytic varieties. Conversely, suppose that V and V' determine equivalent germs of analytic varieties, so that for some representative subvarieties there are analytic mappings $\varphi: V \to V'$ and $\psi: V' \longrightarrow V$ such that the compositions $\psi\circ\varphi: V \longrightarrow V$ and $\varphi\circ\psi: V' \longrightarrow V'$ are the identity mappings; and let Φ and Ψ be extensions of the analytic mappings φ and ψ , respectively, to some open neighborhoods of the origin in $\mathbb{C}^n$. In terms of the coordinates in $\mathbb{C}^n$, write $(\Psi\circ\Phi)(z_1,\ldots,z_n) = (f_1(z),\ldots,f_n(z))$

for some holomorphic functions $f_i(z)$. Since the mapping $\Psi\circ\Phi$ is the identity on the subvariety V , necessarily $f_i(z_1,\ldots,z_n) = z_i$ for any point $(z_1,\ldots,z_n) \in V$, and therefore $f_i(z)-z_i \in \text{id}\, V \subset {}_n\mathcal{O}$; consequently, from Corollary 3 to Theorem 15 it follows that the differential of the function $f_i(z) - z_i$ is zero at the origin, or equivalently, that $\partial f_i(0)/\partial z_j = \delta^i_j$ for the usual Kronecker symbol δ^i_j . This shows that the composition $\Psi\circ\Phi$ is a complex analytic homeomorphism in some open neighborhood of the origin in $\mathbb{C}^n$, so of course the mapping Φ must itself be a complex analytic homeomorphism in some open neighborhood of the origin; this mapping Φ then exhibits V and V' as equivalent germs of analytic subvarieties at the origin in $\mathbb{C}^n$, and the proof is thereby concluded.

Note that as a consequence of Corollary 4, a germ of an analytic variety can be represented by a unique neat germ of an analytic subvariety; for any two neat germs of analytic subvarieties representing the same germ of an analytic variety are necessarily equivalent germs of subvarieties.

The previous considerations extend in part to arbitrary local rings; no attempt will be made here to carry out such extensions in general, but a few remarks should be made in view of some later applications. The imbedding dimension of an arbitrary local ring $\mathcal{O}$ can be defined to be the minimal number of generators of the maximal ideal $\mathfrak{m}$ of that local ring; for the local ring of a germ of an analytic variety this agrees with the geometrical definition of the imbedding dimension, as a consequence of Theorem 15.

The proof of Corollary 1 to Theorem 15 is purely algebraic, so that in general the imbedding dimension of a local ring $\mathcal{O}$ coincides with the dimension of the vector space $\mathfrak{m}/\mathfrak{m}^2$ over the residue class field $\mathcal{O}/\mathfrak{m}$. It is true in general that the Krull dimension of a local ring $\mathcal{O}$ is less than or equal to the imbedding dimension, although there is something to prove here for an arbitrary local ring; a local ring is called a regular local ring if its Krull dimension is equal to its imbedding dimension. The local ring ${}_V\mathcal{O}$ of a germ V of an analytic variety is thus a regular local ring if and only if V is a germ of a regular analytic variety.

§5. The local parametrization theorem for analytic varieties

(a) If V is the germ of an irreducible analytic subvariety at the origin in $\mathbb{C}^n$, and if the coordinates in $\mathbb{C}^n$ are suitably chosen, the natural projection mapping $\mathbb{C}^n \longrightarrow \mathbb{C}^k$ induces a complex analytic mapping $\pi: V \longrightarrow \mathbb{C}^k$ having the various properties listed in the local parametrization theorem. This type of mapping is a useful tool in studying the local properties of complex analytic subvarieties, and can be adapted to be equally useful as a tool in studying the local properties of complex analytic varieties. Indeed, in many ways the local parametrization theorem expressed in terms of complex analytic varieties is easier to use than the version for complex analytic subvarieties discussed in Section 2.

To begin the discussion, consider two germs V_1, V_2 of complex analytic varieties. A continuous mapping $\varphi: V_1 \longrightarrow V_2$ is of course just a continuous mapping between the underlying germs of topological spaces, and determines continuous mappings between any two germs of complex analytic subvarieties representing the given germs of varieties; note that if any one of these latter mappings between germs of subvarieties is analytic, then all the mappings are analytic; and the mapping will be called an <u>analytic mapping</u> between the given germs of analytic varieties. Of course, recalling Theorem 10, a continuous mapping $\varphi: V_1 \longrightarrow V_2$ is analytic if and only if $\varphi^*({}_{V_2}\mathcal{O}) \subseteq {}_{V_1}\mathcal{O}$ for the induced mapping of functions. For the global analogue, consider two complex analytic varieties V_1, V_2. A continuous mapping $\varphi: V_1 \longrightarrow V_2$ is just a continuous mapping between the underlying topological spaces; and such a mapping will be called an analytic mapping if it induces an analytic mapping between the germs of analytic varieties at each point of V_1, or equivalently, if $\varphi^*({}_{V_2}\mathcal{O}_{\varphi(p)}) \subseteq {}_{V_1}\mathcal{O}_p$ for each point $p \in V_1$. In particular, a continuous mapping $\varphi: V \longrightarrow \mathbb{C}^k$ is analytic precisely when the coordinate functions of the mapping φ are holomorphic functions on V. The special analytic mappings which arise in the local parametrization theorem can be described as in the following definition.

A <u>branched analytic covering</u> $\pi: V \longrightarrow U$ is a proper, light, analytic mapping from a complex analytic variety V onto an open subset $U \subseteq \mathbb{C}^k$, such that there exists a complex analytic subvariety $D \subset U$ for which $V - \pi^{-1}(D)$ is dense in V and the restriction

$$\pi: V - \pi^{-1}(D) \longrightarrow U\text{-}D$$

is a complex analytic covering projection. The last condition means that for a sufficiently small open neighborhood $U_z \subseteq U\text{-}D$ of any point $z \in U\text{-}D$, the inverse image $\pi^{-1}(U_z)$ consists of a number of components such that the restriction of π to each component is a complex analytic equivalence between that component and U_z; consequently $V - \pi^{-1}(D)$ is a complex analytic manifold of pure dimension k, and since it is dense in V, necessarily the variety V is itself of pure dimension k. The subset $B = \pi^{-1}(D) \subset V$ will be called the <u>critical locus</u> of the branched analytic covering; it is clear that B is a complex analytic subvariety of V and that $\mathcal{S}(V) \subseteq B$ where $\mathcal{S}(V)$ is the singular locus of the variety V. That branched analytic coverings behave very much like ordinary covering projections is indicated by the following useful auxiliary result.

Localization Lemma. Let $\pi\colon V \longrightarrow U$ be a branched analytic covering; and selecting a point $z \in U$, let $\pi^{-1}(z) = \{p_1,\ldots,p_s\}$ where p_i are distinct points of V. Then there are arbitrarily small open neighborhoods U_z of the point z in U such that $\pi^{-1}(U_z)$ consists of s connected components $V_1,\ldots,V_s$ with $p_i \in V_i$ for each index i; the sets V_i so arising form a basis for the open neighborhoods of the point p_i in the topology of V, and for any such set V_i the restriction $\pi\colon V_i \longrightarrow U_z$ is also a branched analytic covering.

Proof. Selecting any disjoint open neighborhoods V_i' of the points p_i in V, note first that $\pi^{-1}(U_z) \subseteq \cup_i V_i'$ for any sufficiently small open neighborhood U_z of z in U. (For otherwise there would exist a sequence of points $q_j \in V$ such that the image points $\pi(q_j)$ converge to z but $q_j \notin \cup_i V_i'$. Since the points $\pi(q_j)$ together with z form a compact subset in U, and since the mapping π is proper, a subsequence of the points q_j must converge to a limit point $q \in V$; and clearly q must be one of the points p_i, since $\pi(q) = \lim \pi(q_j) = z$. But this is impossible, since $q \notin \cup_i V_i'$.) Now choosing a connected such neighborhood U_z, let V^* be a connected component of $\pi^{-1}(U_z)$ in V;

then $V^* \subseteq V_i'$ for one of the neighborhoods V_i'. Note further that $\pi(V^*) = U_z$. (For since $V^* - V^* \cap B = V^* \cap (V-B)$ is dense in V^* and is a covering space of the connected set $U_z - U_z \cap D$, it is evident at least that $\pi(V^*) \supseteq U_z - U_z \cap D$. If there were a point $z_o \in U_z - \pi(V^*)$, it would necessarily lie in the point set closure $\overline{\pi(V^*)}$ of the set $\pi(V^*)$ in U_z. Select a sequence of points $z_j \in \pi(V^*)$ converging to z_o, and a sequence of points $q_j \in V^*$ such that $\pi(q_j) = z_j$. Again, since the mapping π is proper, a subsequence of the points q_j must converge to a limit point $q_o \in \pi^{-1}(z_o) \cap \overline{V^*} \subseteq \pi^{-1}(U_z) \cap \overline{V^*}$; but this evidently implies that $q_o \in V^*$, which is impossible since $\pi(q_o) = z_o \notin \pi(V^*)$.) Consequently V^* must be an open neighborhood of the point p_i in $V_i' \subseteq V$. That the restriction $\pi: V^* \longrightarrow U_z$ is a branched analytic covering is quite apparent, and the proof is thereby concluded.

As a first consequence of the localization lemma, note that for a branched analytic covering $\pi: V \longrightarrow U$ there are arbitrarily small open neighborhoods of any point $p \in V$ such that the restriction of the mapping π exhibits each neighborhood as a branched analytic covering; therefore the germ of a branched analytic covering is a well defined notion, and it is possible to speak of a germ of an analytic variety as being represented as a germ of a branched analytic covering. There is no loss of generality in assuming that the point p is mapped onto the origin in $\mathbb{C}^k$ under the mapping π, and this normalization of the local version of a branched analytic covering will always be chosen. The mapping π can then be described directly in terms of the germ V by its k coordinate functions, which under the normalization adopted will be k elements of the maximal ideal ${}_V\mathfrak{m} \subset {}_V\mathcal{O}$ of the local ring of the germ V ; any set of k elements of the maximal ideal ${}_V\mathfrak{m}$ of a germ V of an analytic variety of pure dimension k , which arise as the coordinate functions of the germ of a branched analytic covering, will be called a <u>system of parameters</u> for the germ V .

The basic elementary existence and characterization result for branched analytic coverings is the following.

<u>Theorem 16.</u> (a) If V is a complex analytic variety of pure dimension k and $\pi: V \longrightarrow \mathbb{C}^k$ is a complex analytic mapping such that a point $p \in V$ is an isolated point of the subvariety $\pi^{-1}(\pi(p)) \subseteq V$, then the restriction of the mapping π to some open neighborhood of the point p is a branched analytic covering.

(b) If V is the germ of a complex analytic variety of pure dimension k , then a set of k elements of the maximal ideal ${}_V\mathfrak{m} \subset {}_V\mathcal{O}$ form a system of parameters for the germ V if and only if they generate an ideal $\mathfrak{A} \subset {}_V\mathcal{O}$ such that ${}_V\mathfrak{m} = \sqrt{\mathfrak{A}}$.

(c) For any germ $\pi: V \longrightarrow U \subseteq \mathbb{C}^k$ of a branched analytic covering, the germ V can be represented by a complex analytic subvariety V of an open neighborhood of the origin in $\mathbb{C}^n$ such that the coordinates in $\mathbb{C}^n$ are a regular system of coordinates for the ideal $\text{id}\, V \subset {}_n\mathcal{O}$ and the mapping π is induced by the natural projection mapping $\mathbb{C}^n \longrightarrow \mathbb{C}^k$.

Proof. If V is a complex analytic variety of pure dimension k , and $\pi: V \longrightarrow \mathbb{C}^k$ is a complex analytic mapping such that $\pi(p) = 0 \in \mathbb{C}^k$, then p is an isolated point of the subvariety $\pi^{-1}(\pi(p) \subseteq V$ precisely when the coordinate functions of the mapping π generate an ideal $\mathfrak{A} \subset {}_V\mathcal{O}_p$ such that $\operatorname{loc} \mathfrak{A} = p$; and by the Hilbert zero theorem, this is in turn equivalent to the condition that $\sqrt{\mathfrak{A}} = \operatorname{id} p = {}_V\mathfrak{M} \subset {}_V\mathcal{O}_p$. Therefore parts (a) and (b) are really equivalent, and for their proof it is only necessary to show that any set of k germs $f_1,\dots,f_k$ in ${}_V\mathcal{O}_p$ generating an ideal $\mathfrak{A} \subset {}_V\mathcal{O}_p$ with $\operatorname{loc} \mathfrak{A} = p$ is a system of parameters for the germ V . Suppose therefore that those germs are represented by analytic functions $f_1,\dots,f_k$ on a complex analytic subvariety V of an open neighborhood U^* of the origin in $\mathbb{C}^n$, where the subvariety V represents the given germ of a variety; and suppose that the set of common zeros of these functions consists of the origin alone, the origin being the point p . If the neighborhood U^* is sufficiently small, the irreducible components of the germ V will be represented by separate analytic subvarieties of U^* and the functions $f_1,\dots,f_k$ will be the restrictions to V of analytic functions $F_1,\dots,F_k$ in U^* . In terms of the coordinates $z_1,\dots,z_n$ in $\mathbb{C}^n$ introduce the complex analytic mapping $(z_1,\dots,z_n) \longrightarrow (F_1(z),\dots,F_k(z),z_1,\dots,z_n)$ from U^* into $\mathbb{C}^{k+n}$; this is clearly a complex analytic homeomorphism from U^* onto a complex analytic submanifold of $\mathbb{C}^k \times U^* \subseteq \mathbb{C}^n$, hence the image of the subvariety $V \subseteq U^*$ is a complex analytic subvariety $V_1 \subseteq \mathbb{C}^k \times U^*$ representing the same germ of a complex analytic variety as V. For this representative, the functions $f_1,\ldots,f_k$ are the restrictions to the subvariety V_1 of the first k coordinates $w_1,\ldots,w_k$ in the ambient space $\mathbb{C}^{k+n}$. Now since by assumption $V_1 \cap \{w | w_1 = \ldots = w_k = 0\}$ consists of the origin alone, it follows from Theorem 8(b) that the coordinates $w_1,\ldots,w_n$ form a regular system of coordinates for the prime ideals in ${}_{k+n}\mathcal{O}$ corresponding to the various irreducible components of the germ of V_1 at the origin; hence from the local parametrization theorem it further follows that the natural projection $\mathbb{C}^{k+n} \longrightarrow \mathbb{C}^k$ exhibits each irreducible component of V_1 as a branched analytic covering of some open neighborhood of the origin in $\mathbb{C}^k$. This projection is the analytic mapping defined by the given analytic functions $f_1,\ldots,f_k$, and the entire subvariety is then represented as a branched analytic covering of an open neighborhood of the origin in $\mathbb{C}^k$ by this mapping; therefore $f_1,\ldots,f_k$ form a system of parameters for the analytic variety V, as desired. Actually, part (c) has been proved at the same time, so that the entire proof is concluded.

One special case of this theorem is perhaps worth discussing separately. Any n functions which are holomorphic in an open neighborhood V of the origin in $\mathbb{C}^n$ and which vanish at the origin determine an analytic mapping $\pi: V \longrightarrow \mathbb{C}^n$; and this mapping is a branched analytic covering at the origin in V if and only if the origin is an isolated point of the subvariety of common zeros of these n functions. For $n = 1$, the set of zeros of a non-constant analytic function is always isolated, so that any non-constant function of one complex variable defines a branched analytic covering. This is of course a very familiar result; and it is further familiar that the standard form of a branched analytic covering, in some sense, is the mapping $z \longrightarrow z^r$. An immediate corollary is that any non-constant analytic function of a single complex variable determines an open mapping. For $n > 1$, the situation is rather more complicated, and branched analytic coverings are but a special type of non-trivial analytic mapping. Although it follows immediately from the localization lemma that a branched analytic covering is an open mapping, there are non-trivial mappings which are not open, even though their images have non-empty interiors. The simplest example is probably the analytic mapping $(z_1,z_2) \longrightarrow (z_1,z_1z_2)$; the image of this mapping is the complement of the set $\{(z_1,z_2) \in \mathbb{C}^2 | z_1 = 0, z_2 \neq 0\}$, hence the mapping is evidently not open.

(b) If $\pi: V \longrightarrow U$ is a branched analytic covering, the restriction of the mapping π to the complement of the critical locus is necessarily a finite-sheeted covering mapping; the number of sheets will be called the <u>order</u> of the branched analytic covering. For any point $p \in V$, it follows from the localization lemma that there are arbitrarily small open neighborhoods V_p of

p in V such that the restriction of the mapping π to V_p is also a branched analytic covering. Since the orders of these local branched analytic coverings can only decrease as the neighborhoods V_p shrink to the point p, it is evident that the order is the same for all sufficiently small such neighborhoods; this common order will be called the branching order of the mapping π at the point p, and will be denoted by $o(p)$. Note that if $\pi: V \longrightarrow U$ is a branched analytic covering of order r, then selecting any point $z \in U$ and letting $p_1, \ldots, p_r$ be the distinct points of the subset $\pi^{-1}(z) \subseteq V$, it follows again from the localization lemma that $\Sigma_i \, o(p_i) = r$. A useful convention is to list a point $p \in \pi^{-1}(z)$ a total of $o(p)$ times; then $\pi^{-1}(z)$ is a set consisting of r elements for each point $z \in U$, but the elements are not necessarily distinct points of V.

For any point p outside the critical locus B of a branched analytic covering $\pi: V \longrightarrow U$, it is of course clear that $o(p) = 1$; however there may very well be points p in the critical locus for which $o(p) = 1$, since not all the points of $\pi^{-1}(\pi(p))$ need necessarily have the same branching order, even when the critical locus is chosen to be as small as possible. There is thus some point to introducing the subset $B_o = \{p \in V | o(p) > 1\} \subseteq B \subset V$, which will be called the branch locus of the branched analytic covering $\pi: V \longrightarrow U$; points in the branch locus will be called branch points of that branched analytic covering.

<u>Theorem 17.</u> The branch locus B_o of a branched analytic covering $\pi: V \longrightarrow U$ is an analytic subvariety of the analytic variety V. The intersection of the germs of the branch loci, for all representations of the germ V of an analytic variety as a branched analytic covering, is precisely the singular locus $\mathcal{S}(V)$ of the germ V.

Proof. The entire theorem is really of a local nature. Hence it can be assumed that the variety V is represented by a complex analytic subvariety V of an open subset $U^* \subseteq \mathbf{C}^n$, and that as in Theorem 7 there are holomorphic functions $f_1,\ldots,f_m$ in U^* which generate the ideal $\text{id}\, V \subset {}_n\mathcal{O}_a$ at each point $a \in U^*$; and further, it can be assumed that there are holomorphic functions $g_1,\ldots,g_k$ in U^* whose restrictions to the subvariety V are the coordinate functions for the mapping π. Note that at any point $a \in U^*$ the differentials $df_i(a)$ and $dg_j(a)$ can be viewed as vectors in $\mathbf{C}^n$. The first step in the proof is then to show that

$$B_o = \{a \in V \subseteq U^* | \operatorname{rank}(df_1(a),\ldots,df_m(a),\ dg_1(a),\ldots,dg_k(a)) < n\} .$$

To do this, consider a point $a \in V$ at which the above rank is equal to n. In an open neighborhood of that point a, some n of the functions $f_1,\ldots,f_m,g_1,\ldots,g_k$ can be taken as local coordinate functions in $\mathbf{C}^n$; clearly these n functions must include all of the functions $g_1,\ldots,g_k$, since the set of common zeros of all of the functions $f_1,\ldots,f_m$ and any $k-1$ of the functions $g_1,\ldots,g_k$ is an analytic subvariety of positive dimension and not just a point. It follows immediately that V is a complex analytic manifold of dimension k in a neighborhood of the point a, and that the functions $g_1,\ldots,g_k$ are a regular system of parameters there, so that $o(p) = 1$. Next consider a point $a \in V$ at which $o(a) = 1$. The mapping defined by the functions $g_1,\ldots,g_k$ is then a topological homeomorphism between a neighborhood V_a of the point a in V and an open subset $U_a \subseteq \mathbb{C}^k$. The inverse homeomorphism is described by n continuous functions $h_1,\ldots,h_n$ in U_a, such that $(h_1(w),\ldots,h_n(w)) \in V_a$ for any point $w \in U_a$; these functions are holomorphic in $U_a - U_a \cap D$, hence by the Riemann removable singularities theorem are holomorphic throughout U_a, so that V_a is actually a regular analytic variety. Thus V_a can be defined by the vanishing of $n-k$ of the functions $f_1,\ldots,f_m$, say by $f_1,\ldots,f_{n-k}$; and $\operatorname{rank}(df_1(a),\ldots,df_{n-k}(a)) = n-k$. Moreover, on the k-dimensional manifold defined by the vanishing of the functions $f_1,\ldots,f_{n-k}$, the mapping defined by the functions $g_1,\ldots,g_k$ is a complex analytic homeomorphism; and from this it is easy to see that $\operatorname{rank}(df_1(a),\ldots,df_{n-k}(a), dg_1(a),\ldots,dg_k(a)) = n$.

Having established this preliminary result, it is obvious that B_o is an analytic subvariety of V; for B_o consists of those points of V at which all the $n \times n$ determinants formed from the matrix of holomorphic functions $(df_1(z),\ldots,df_m(z), dg_1(z),\ldots,dg_k(z))$ are zero. Now consider the germ of the subvariety V at the origin in $U^* \subseteq \mathbb{C}^n$, and assume as usual that the system of parameters $g_1,\ldots,g_k$ take the origin in $\mathbb{C}^n$ to the origin in $\mathbb{C}^k$. For any complex constant matrix $c = (c_{ij})$, $1 \leq i \leq k$, $1 \leq j \leq n$, consider also the functions $g_i^c(z) = g_i(z) + \Sigma_j\, c_{ij} z_j$, which are holomorphic in U^* and also vanish at the origin in $\mathbb{C}^n$. It follows immediately from the semicontinuity lemma that, whenever $|c_{ij}| < \varepsilon$ for a suitably small $\varepsilon > 0$, the set of common zeros of the functions $g_1^c, \ldots, g_k^c$ on V has the origin as an isolated point; consequently, applying Theorem 16, whenever $|c_j| < \varepsilon$ the functions $g_1^c, \ldots, g_k^c$ also form a system of parameters for the germ V of an analytic variety. If as germs at the origin in $\mathbb{C}^n$, $\mathcal{S}(V) \subset B_o$, where B_o is the branch locus of the branched analytic covering defined by the system of parameters $g_1, \ldots, g_k$, select a point a lying in some component of B_o passing through the origin but not lying in the singular set $\mathcal{S}(V)$. Then it follows as in Theorem 12 that $\operatorname{rank}(df_1(a), \ldots, df_m(a)) = n-k$. Note though that $dg_i^c(a) = dg_i(a) + c_i$ for the constant vector $c_i = (c_{ij})$; consequently there exist constants $|c_{ij}| < \varepsilon$ such that $\operatorname{rank}(df_1(a), \ldots, df_m(a),\ dg_1^c(a), \ldots, dg_k^c(a)) = n$. This shows that the point a does not lie in the branch locus B_o^c of the branched analytic covering defined by the system of parameters $g_1^c, \ldots, g_k^c$, and therefore necessarily $B_o \cap B_o^c \subset B_o$ as germs at the origin in $\mathbb{C}^n$. If $\mathcal{S}(V) \subset B_o \cap B_o^c$, the process can be repeated, yielding a system of parameters $g_1^{c'}, \ldots, g_k^{c'}$ with branch locus $B_o^{c'}$ such that $B_o \cap B_o^c \cap B_o^{c'} \subset B_o \cap B_o^c$ as germs at the origin in $\mathbb{C}^n$.

Since the ring ${}_V\mathcal{O}$ is Noetherian this process must terminate after a finite number of repetitions, and then $\mathcal{S}(V) = B_o \cap B_o^{c'} \cap B_o^{c'} \cap \ldots$, which suffices to complete the proof.

It follows from the proof of the preceding theorem, or more generally from the Noetherian property of the ring ${}_V\mathcal{O}$ together with the statement of the preceding theorem, that the singular locus $\mathcal{S}(V)$ can be written as the intersection of finitely many branch loci, in the representations of the germ V of an analytic variety as a branched analytic covering in various ways. Furthermore, for any particular representation of the germ V as a branched analytic covering, $\mathcal{S}(V) \subseteq B_o \subseteq B$; and these can be proper containment relations. Indeed, a complex analytic manifold V can be represented as a branched analytic covering with a non-trivial branch locus. On the other hand, whenever $p \in V - B_o$ then necessarily $p \in \mathcal{R}(V)$. A system of parameters for the germ of the analytic variety V at a point $p \in V$ will be called a <u>regular system of parameters</u> if $o(p) = 1$ in the branched analytic covering they define, or equivalently, if the germ of a branched analytic covering they define has an empty branch locus; the point p must then be a regular point of the variety V. Note that a regular system of parameters for the germ V of an analytic variety actually form a set of coordinates on the manifold V, hence they generate the full maximal ideal ${}_V\mathcal{M} \subset {}_V\mathcal{O}$ rather than just an ideal having ${}_V\mathcal{M}$ as its radical. Conversely of course, whenever a system of parameters generate the full maximal ideal ${}_V\mathfrak{m} \subset {}_V\mathcal{O}$, then imbed dim V = dim V , so that V is locally a manifold; and the system of parameters form local coordinates, hence are a regular system of parameters. Thus a system of parameters is regular if and only if they generate the full maximal ideal; and a local ring ${}_V\mathcal{O}$ admits a regular system of parameters if and only if the germ V of analytic variety is regular.

Branch points of two quite different kinds can occur in a branched analytic covering $\pi: V \rightarrow U$. A point $p \in B_o$ will be called an <u>essential branch point</u> if there exist arbitrarily small open neighborhoods V_p of the point p in the variety V such that $V_p - V_p \cap B_o$ is a connected set. The remaining points of B_o will be called the <u>accidental branch points</u>; thus for all sufficiently small open neighborhoods V_p of an accidental branch point in a variety V , the set $V_p - V_p \cap B_o$ has at least two connected components. Note that if $p \in \mathcal{R}(V) \cap B_o$ then p is necessarily an essential branch point; for any sufficiently small connected open neighborhood V_p of the point p is a connected complex analytic manifold, so the complement of the analytic subvariety $V_p \cap B_o \subset V_p$ is necessarily connected. This observation can be rephrased upon application of Theorem 17, in an apparently more confusing manner, as the assertion that those branch points of a branched analytic covering which cease to be branch points in some other representations of the variety as a branched analytic covering are necessarily essential branch points; this is really not confusing, if it is remarked that the property that a branch point be essential is really a property of the branched analytic covering and is not intrinsic to the complex analytic variety so represented. Representing the variety V in some neighborhood of the point $p \in V$ by an analytic subvariety of an open subset of $\mathbb{C}^n$ such that the branched analytic covering $\pi: V \longrightarrow U$ is induced by the natural projection mapping $\mathbb{C}^n \longrightarrow \mathbb{C}^k$ as in the local parametrization theorem, it follows from Corollary 4 to Theorem 5 that the separate components of $V_p - V_p \cap B_o$, for sufficiently small connected open neighborhoods V_p of the point p in V, correspond to separate irreducible components of the germ of the variety V at the point p. Thus the accidental branch points of the branched analytic covering, which necessarily lie in the singular locus $\mathcal{S}(V)$ of the variety V, are precisely those points at which the germ of the variety is irreducible. If the variety V is irreducible at each point, then all the branch points in any representation of V as a branched analytic covering are essential branch points. It would be natural to attempt to pull apart all the accidental branch points, to obtain an analytic variety more primitive than the original variety, such that the original variety arises by imbedding that more primitive variety in such a manner that some accidental intersections arise; but this cannot be done too brutally, since accidental branch points may have a limit point which is an essential branch point. For example, the analytic subvariety $V \subset \mathbb{C}^3$ defined by the single equation $z_3^2 - z_2^2 z_1 = 0$ is exhibited as a branched analytic covering of $\mathbb{C}^2$

under the natural projection mapping $\mathbb{C}^3 \longrightarrow \mathbb{C}^2$, and the branch locus is the subvariety $B_o = \{(z_1,z_2,z_3) \in V | z_1 z_2 = 0\}$; the points $(z_1,0,0)$ for $z_1 \neq 0$ are accidental branch points of V , while the points $(0,z_2,0)$ for all z_2 are essential branch points of V. As noted earlier, the germ V is irreducible at the origin in $\mathbb{C}^3$, but not at any point of the form $(z_1,0,0)$ for $z_1 \neq 0$.

(c) The canonical equations for a prime ideal $\mathfrak{p} \subset {}_n\mathcal{O}$ played a fundamental role in the derivation of the local parametrization theorem; and it might be anticipated that their analogues for branched analytic coverings would have a comparable use. In this latter case, though, the geometric properties can be taken to some extent as being given; but the analogues of the canonical equations provide useful tools in deriving further algebraic and analytic properties. The advantage of this reversal in point of view is that a comparison of the discussion here with that in §2 may perhaps clarify the earlier considerations, by exhibiting more explicitly the analytic structure underlying the previous almost purely algebraic constructions.

First, as a preliminary observation, if $\pi: V \longrightarrow U$ is a branched analytic covering and h is a holomorphic function on all of $U \subseteq \mathbb{C}^k$, then the composition $\pi^*(h) = h \circ \pi$ is a holomorphic function on all of V ; and this mapping $\pi^*: {}_k\mathcal{O}_U \longrightarrow {}_V\mathcal{O}_V$ is clearly an isomorphism from the ring of holomorphic functions on U into the ring of holomorphic functions on V , since the image of π is all of U . It is convenient to identify the ring ${}_k\mathcal{O}_U$ with its isomorphic image $\pi^*({}_k\mathcal{O}_U)$, as a subring of ${}_V\mathcal{O}_V$; and with this convention, the characteristic properties of the canonical equations can be summarized as follows.

Theorem 18. Let $\pi: V \longrightarrow U$ be a branched analytic covering of order r over an open subset $U \subseteq \mathbb{C}^k$.

(a) For any analytic function $f \in {}_V\mathcal{O}_V$ there exists a unique monic polynomial $p_f(X) \in {}_k\mathcal{O}_U[X]$ of degree r such that $p_f(f) = 0$ on V.

(b) If $g \in {}_V\mathcal{O}_V$ is an analytic function on V such that the discriminant $d_g \in {}_k\mathcal{O}_U$ of the polynomial $p_g(X) \in {}_k\mathcal{O}_U[X]$ is not identically zero, then for any analytic function $f \in {}_V\mathcal{O}_V$ there exists a unique polynomial $q_{f,g}(X) \in {}_k\mathcal{O}_U[X]$ of degree $r-1$ such that $d_g \cdot f = q_{f,g}(g)$ on V.

Proof. (a) The polynomial $p_g(X)$ is constructed exactly as in the proof of the continuation of Theorem 5. For any point $z \in U$ let $\pi^{-1}(z) = \{p_1(z),\ldots,p_r(z)\} \subseteq V$, where a point is repeated according to its branching order so that there are always precisely r points of this set but they are not necessarily always distinct points; it is evident that the only possible such polynomial $p_f(X)$ must be

$$p_f(z,X) = \prod_{i=1}^{r} (X - f(p_i(z))) ,$$

and it is only necessary to show that the coefficients of this polynomial are analytic functions in U. Note that these coefficients are the elementary symmetric functions of the values $f(p_i(z))$, hence are independent of the order in which the points of the set $\pi^{-1}(z)$ are written. Now it follows quite trivially from the localization lemma that the coefficients are continuous functions of z, since the function f is itself continuous on V. Furthermore, in an open neighborhood of any point $z_o \in U$ outside of the image D of the critical locus, the branched analytic covering is an r-sheeted analytic covering in the ordinary sense; hence in that neighborhood it is possible so to label the points of the set $\pi^{-1}(z)$ that each $p_i(z)$ is a complex analytic mapping from a neighborhood of z_o in U into V. Thus the coefficients are clearly analytic functions in $U-D$, and since they are continuous in U it follows from the Riemann removable singularities theorem that they are analytic throughout U, as desired.

(b) Consider next an analytic function $g \in {}_V\mathcal{O}_V$ such that the discriminant $d_g \in {}_k\mathcal{O}_U$ of the polynomial $p_g(X) \in {}_k\mathcal{O}_U[X]$ is not identically zero in U. Retaining the notation adopted in the first part of the proof, recall that the discriminant is given by $d_g(z) = \prod_{i \neq j} [g(p_i(z)) - g(p_j(z))]$; thus the condition on the discriminant is equivalent to the condition that the r values $g(p_i(z))$, $i = 1,\ldots,r$, are distinct for at least one point $z \in U$. Now for any other analytic function $f \in {}_V\mathcal{O}_V$ the problem is to find analytic functions $a_j \in {}_k\mathcal{O}_U$ such that

$$(*) \qquad d_g(z)\cdot f(p_i(z)) = \sum_{j=o}^{r-1} a_j(z)\cdot g(p_i(z))^j, \quad i = 1,\ldots,r,$$

for any point $z \in U$; for then the polynomial

$$q_{f,g}(z,X) = \sum_{j=o}^{r-1} a_j(z)X^j$$

has the desired properties. Note that the equations (*) can be viewed as a system of r linear equations in the r unknown values $a_j(z)$, at any fixed point $z \in U$; hence by Cramer's rule

$$a_j(z)\cdot\det[1,\ g(p_i(z)),\ g(p_i(z))^2,\ldots,g(p_i(z))^{r-1}]$$

$$= \det[1,g(p_i(z)),\ldots,g(p_i(z))^{j-1},d_g(z)f(p_i(z)),g(p_i(z))^{j+1},\ldots,g(p_i(z))^{r-1}],$$

where in both determinants the entries in row i are as indicated. The determinant appearing in the left hand side of the last equation is the van der Monde determinant Δ , and it is well known that $\Delta^2 = d_g(z)$ is the discriminant of the polynomial $p_g(X)$ at the point z . If $d_g(z) \neq 0$, then factoring that term from the determinant on the right hand side of the last equation and dividing by Δ on both sides produces the explicit formula

$$a_j(z) = \Delta\cdot\det[1,g(p_i(z)),\ldots,g(p_i(z))^{j-1},f(p_i(z)),g(p_i(z))^{j+1},\ldots,g(p_i(z))^{r-1}] .$$

Note that both determinants on the right hand side change sign upon interchanging any two rows; the product is thus invariant under the simultaneous change of any pair of rows in the two factors, hence is really independent of the order in which the points of the set $\pi^{-1}(z)$ are labelled. Again it follows trivially from the localization lemma that the functions $a_j(z)$ defined above are continuous in all of U , and they are analytic in $U-D$ since locally the functions $p_i(z)$ can be chosen to be analytic mappings; so by the Riemann removable singularities theorem the functions $a_j(z)$ are analytic throughout U , and the proof is thereby concluded.

If V is the germ at the origin in $\mathbb{C}^n$ of an irreducible k-dimensional analytic subvariety, and if the coordinates $z_1,\dots z_n$ in $\mathbb{C}^n$ are chosen to be strictly regular for the ideal $\text{id}\, V \subset {}_n\mathcal{O}$, then the natural projection $\mathbb{C}^n \longrightarrow \mathbb{C}^k$ induces a branched analytic covering $\pi\colon V \longrightarrow U$. The restrictions to the subvariety V of the coordinate functions $z_{k+1},\dots,z_n$ are complex analytic functions on V , and the polynomials $p_{z_j}(X) \in {}_k\mathcal{O}_U[X]$ of Theorem 18(a) lead to the first set of canonical equations $p_j = p_{z_j}(z_j)$ for the ideal $\text{id}\, V \subset {}_n\mathcal{O}$; and letting $d \in {}_k\mathcal{O}$ be the discriminant of the canonical equation $p_{k+1} = p_{z_{k+1}}$, the polynomials $q_{z_j,z_{k+1}}(X) \in {}_k\mathcal{O}_U[X]$ of Theorem 18(b) lead to the second set of canonical equations $q_j = d\cdot z_j - q_{z_j,z_{k+1}}(z_{k+1})$ for the ideal $\text{id}\, V \subset {}_n\mathcal{O}$. It is in this sense that Theorem 18 can be considered as extending the canonical equations of analytic subvarieties to branched analytic coverings. The canonical equations were only established for prime ideals in §2; but clearly Theorem 18 can be used to derive the canonical equations for an arbitrary pure-dimensional ideal in the ring ${}_n\mathcal{O}$.

Note that the condition that the discriminant $d_g \in {}_k\mathcal{O}_U$ of the polynomial $p_g(X) \in {}_k\mathcal{O}_U[X]$ not be identically zero is equivalent to the condition that the values $g(p_j(z))$ be distinct for at least one point $z \in U$, as remarked in the proof of the theorem. Actually of course these values are distinct for all points $z \in U$ for which $d_g(z) \neq 0$; so in this sense, the condition is that the values of the function g generally separate the sheets of the branched analytic covering $\pi\colon V \longrightarrow U$.

(d) As noted before, if $\pi\colon V \longrightarrow U$ is a branched analytic covering and f is a holomorphic function on the subset $U \subseteq \mathbb{C}^k$, then the composition $\pi^*(f) = f\circ\pi$ is a holomorphic function on the variety V; and the mapping $\pi^*\colon {}_k\mathcal{O}_U \longrightarrow {}_V\mathcal{O}_V$ so defined is clearly an isomorphism from the ring of holomorphic functions on U into the ring of holomorphic functions on V. For any point $p \in V$, the localization lemma shows that there are arbitrarily small open neighborhoods of the point p such that the restrictions of π to these neighborhoods are also branched analytic coverings; and hence there results the natural local isomorphism $\pi^*\colon {}_k\mathcal{O}_{\pi(p)} \longrightarrow {}_V\mathcal{O}_p$. The ring ${}_V\mathcal{O}_p$ can be viewed as a module over the subring ${}_k\mathcal{O}_{\pi(p)} \cong \pi^*({}_k\mathcal{O}_{\pi(p)}) \subseteq {}_V\mathcal{O}_p$; indeed, it follows readily from the local parametrization theorem that ${}_V\mathcal{O}_p$ is a finitely generated integral algebraic extension over this subring. Actually somewhat more can be shown; viewing the rings

${}_k\mathcal{O}_{\pi(p)} \cong \pi^*({}_k\mathcal{O}_{\pi(p)})$ as forming a subsheaf of rings of the sheaf of rings ${}_V\mathcal{O}$ over the variety V , the sheaf ${}_V\mathcal{O}$ is locally a finitely generated sheaf of modules over this subsheaf of rings $\pi^*({}_U\mathcal{O}) \subseteq {}_V\mathcal{O}$. This is rather reminiscent of the coherence conditions discussed earlier, but is in a sense somewhat topsy-turvy; it is more natural to consider locally finitely generated sheaves of modules over either the structure sheaf ${}_U\mathcal{O}$ of U or the structure sheaf ${}_V\mathcal{O}$ of V , whereas here the structure sheaf ${}_V\mathcal{O}$ is viewed as a locally finitely generated sheaf of modules over a new sheaf of rings $\pi^*({}_U\mathcal{O})$ over V . This can be reversed, to yield a more convenient way of looking at the same situation, by considering the direct image $\pi_*({}_V\mathcal{O})$ of the sheaf of rings ${}_V\mathcal{O}$, rather than the inverse image $\pi^*({}_U\mathcal{O})$ of the sheaf of rings ${}_U\mathcal{O}$. It is perhaps clearer to discuss the relevant sheaf construction somewhat more generally at first, and then to specialize to the case of present interest.

Suppose therefore that $\pi: V \longrightarrow U$ is a continuous mapping between two topological spaces V and U , and that $\mathcal{S}$ is a sheaf of rings (or of groups, etc.) over the space V . To each open set U_α of a basis for the open sets in the topology of U associate the ring $\mathcal{R}_\alpha = \Gamma(\pi^{-1}(U_\alpha), \mathcal{S})$. It is clear that whenever $U_\alpha \subseteq U_\beta$, the natural restriction mapping of a section of $\mathcal{S}$ over $\pi^{-1}(U_\beta)$ to the subset $\pi^{-1}(\alpha) \subseteq \pi^{-1}(U_\beta)$ induces a homomorphism $\rho_{\alpha\beta}: \mathcal{R}_\beta \longrightarrow \mathcal{R}_\alpha$; and that these homomorphisms satisfy $\rho_{\alpha\beta}\rho_{\beta\gamma} = \rho_{\alpha\gamma}$ whenever $U_\alpha \subseteq U_\beta \subseteq U_\gamma$. Thus $\{U_\alpha, \mathcal{R}_\alpha, \rho_{\alpha\beta}\}$ is a presheaf of rings over U ; the associated sheaf will be called <u>the direct image of the sheaf</u> $\mathcal{S}$ <u>under the mapping</u> π , and will be denoted by $\pi_*(\mathcal{S})$. It is obvious that the presheaf just constructed is a complete presheaf, so that the natural homomorphism $\mathcal{R}_\alpha \longrightarrow \Gamma(U_\alpha, \pi_*(\mathcal{S}))$ is an isomorphism; that is to say, $\Gamma(U_\alpha, \pi_*(\mathcal{S})) \cong \Gamma(\pi^{-1}(U_\alpha), \mathcal{S})$ under the obvious canonical isomorphism.

Now for a branched analytic covering $\pi: V \longrightarrow U$, the direct image $\pi_*({}_V\mathcal{O})$ is a sheaf of rings over the open subset $U \subseteq \mathbb{C}^k$. There is moreover a natural isomorphism from ${}_U\mathcal{O}$ into $\pi_*({}_V\mathcal{O})$; for to any section $f_\alpha \in \Gamma(U_\alpha, {}_U\mathcal{O})$ associate the section $\pi^*(f_\alpha) = f_\alpha \circ \pi \in \Gamma(\pi^{-1}(U_\alpha), {}_V\mathcal{O})$, to define a homomorphism from the presheaf of sections of ${}_U\mathcal{O}$ into the presheaf used to define the direct image sheaf $\pi_*({}_V\mathcal{O})$, and observe that this clearly yields an isomorphism from the sheaf ${}_U\mathcal{O}$ into the sheaf $\pi_*({}_V\mathcal{O})$. Under this isomorphism, the sheaf $\pi_*({}_V\mathcal{O})$ can be viewed as a sheaf of modules over the sheaf of rings ${}_U\mathcal{O}$, that is to say, as an analytic sheaf over the open subset $U \subseteq \mathbb{C}^k$.

<u>Theorem 19(a).</u> For any branched analytic covering $\pi: V \longrightarrow U$, the direct image sheaf $\pi_*({}_V\mathcal{O})$ is locally a finitely generated analytic sheaf over U .

Proof. As a consequence of Theorem 16, an open neighborhood of any point on the analytic variety V can be represented by an analytic subvariety V of an open neighborhood of the origin in $\mathbb{C}^n$, such that the mapping π is induced by the natural projection mapping $\mathbb{C}^n \longrightarrow \mathbb{C}^k$ as in the local parametrization theorem. It suffices to prove the theorem for just this piece of the branched analytic covering; for the part of V over any point of U can be written as a disjoint union of such pieces, by the localization lemma, and clearly the direct image sheaf is the direct sum of the direct images of each separate component. If this piece of the branched analytic covering is of order r, consider the monomials $h_\nu = \tilde{z}_{k+1}^{\nu_{k+1}} \cdots \tilde{z}_n^{\nu_n} \in \Gamma(V, {}_V\mathcal{O})$ for $0 \leq \nu_{k+1}, \ldots, \nu_n \leq r-1$, where as before the analytic functions $\tilde{z}_j$ are the restrictions to the subvariety V of the coordinate functions z_j in $\mathbb{C}^n$. The sections $h_\nu \in \Gamma(V, {}_V\mathcal{O})$ induce sections $H_\nu \in \Gamma(U, \pi_*({}_V\mathcal{O}))$; and the proof will be concluded by showing that these sections H_ν generate the sheaf $\pi_*({}_V\mathcal{O})$ as a sheaf of modules over the sheaf of rings ${}_U\mathcal{O}$.

For any point $a \in U$ let $\pi^{-1}(a) = \{p_1, \ldots, p_s\} \subset V$, where p_j are distinct points; and applying the localization lemma, choose an open neighborhood U_a of the point a such that $\pi^{-1}(U_a) \subseteq V$ has s connected components $V_1, \ldots, V_s$ for which $p_j \in V_j$ and the restriction $\pi\colon V_j \longrightarrow U_a$ is a branched analytic covering of order $o(p_i)$. Since the points p_j are distinct and $\pi(p_j) = a$, there are constants $c_{k+1}, \ldots, c_n$ such that the function $g = c_{k+1}z_{k+1} + \ldots + c_n z_n$ takes distinct values at distinct points p_j ; indeed, if the neighborhood U_a is chosen sufficiently small, the function g takes distinct values on the separate components V_j. Now for each component V_j there exists by Theorem 18 a monic polynomial $p_{g,j}(X) \in {}_k\mathcal{O}_{U_a}[X]$ of degree $o(p_j)$ such that $p_{g,j}(g) = 0$ on the component V_j ; since the roots of the polynomial $p_{g,j}(X)$ for any fixed point $z' = (z_1,\ldots,z_k) \in U_a$ are precisely the values of the function g at the points of $\pi^{-1}(z')$ lying in V_j , and since the function g takes distinct values on the separate components of $\pi^{-1}(U_a)$, it further follows that the function $p_{g,j}(g)$ is nowhere zero on the components V_i for $i \neq j$. Note that the restriction g_j of the function $p_{g,j}(g)$ to the analytic subvariety $\pi^{-1}(U_a) = \cup V_i$ is thus a polynomial in $\tilde{z}_{k+1},\ldots,\tilde{z}_n$ of degree $o(p_j)$, which vanishes identically on V_j and is nowhere zero on V_i for $i \neq j$.

Now consider an arbitrary element $F \in \pi_*({}_V\mathcal{O})_a$; recalling the definition of the direct image sheaf, the element F is evidently described by s germs $f_i \in {}_{V_i}\mathcal{O}_{p_i}$, represented by holomorphic functions f_i on the various components V_i of $\pi^{-1}(U_a)$ for a sufficiently small open neighborhood U_a of the point a . The function $f_i / \prod_{j\neq i} g_j$ is analytic on V_i in an open neighborhood of p_i ; consequently, as in the local parametrization theorem, this function on V_i can be written as a polynomial $f_i^* \in {}_k\mathcal{O}_a[\tilde{z}_{k+1},\ldots,\tilde{z}_n]$ of degree at most $o(p_i) - 1$ in each variable. Then $f_i^* \cdot \prod_{j\neq i} g_j \in {}_k\mathcal{O}_a[\tilde{z}_{k+1},\ldots,\tilde{z}_n]$ is a polynomial of degree at most $o(p_i) - 1 + \sum_{j\neq i} o(p_j) = r-1$ in each variable, which agrees with the function f_i on the component V_i and vanishes on the components V_j for $j \neq i$; and consequently $F = \sum_i f_i^* \cdot \prod_{j\neq i} g_j$ is expressed as a linear combination of the elements H_ν with coefficients in ${}_k\mathcal{O}_a$, thus concluding the proof of the theorem.

The construction in the proof of the preceding theorem shows that when the analytic variety V is represented by an analytic subvariety V of an open neighborhood of the origin in $\mathbf{C}^n$, such that the natural projection $\mathbf{C}^n \longrightarrow \mathbf{C}^k$ exhibits V as a branched analytic covering of an open neighborhood U of the origin in $\mathbf{C}^k$, there is an exact sequence of analytic sheaves over U of the form

$$0 \longrightarrow \mathcal{K} \longrightarrow {}_U\mathcal{O}^R \longrightarrow \pi_*({}_V\mathcal{O}) \longrightarrow 0 ;$$

here ${}_U\mathcal{O}^R$ can be identified with the subsheaf ${}_U\mathcal{O}^R \subset {}_U\mathcal{O}[z_{k+1},\ldots,z_n]$ consisting of polynomials of degree at most $r-1$ in each variable, so that $R = r^{n-k}$, and the mapping onto $\pi_*({}_V\mathcal{O})$ can be identified with the restriction of these polynomials to the analytic subvariety $V \subset U \times \mathbf{C}^{n-k}$. Consequently the kernel $\mathcal{K}$ can be identified with the subsheaf $\mathcal{K} \subset {}_U\mathcal{O}^R \subset {}_U\mathcal{O}[z_{k+1},\ldots,z_n]$ consisting of those polynomials vanishing on the analytic subvariety $V \subset U \times \mathbf{C}^{n-k}$. More accurately, for any point $a \in U$ the stalk $\mathcal{K}_a$ consists of those polynomials in ${}_U\mathcal{O}_a[z_{k+1},\ldots,z_n]$ of degree at most $r-1$ in each variable which vanish at all points of the subvariety $V \subset U \times \mathbf{C}^{n-k}$ lying over some sufficiently small open neighborhood U_a of the point a in U . In the special case that $n = k+1$ it is easy to see that $\mathcal{K} = 0$, and consequently that $\pi_*({}_V\mathcal{O}) \cong {}_U\mathcal{O}^R$. To see this, note that any polynomial

$f \in {}_k\mathcal{O}_a[z_{k+1}]$ of degree at most $r-1$, which vanishes on the analytic subvariety $V \cap (U_a \times \mathbb{C}^1)$ for some open neighborhood U_a of the point a, must have r distinct zeros over a dense open subset of U_a, and hence clearly vanishes identically. This is of sufficient interest to merit restating explicitly, as follows.

Corollary 1 to Theorem 19. If $\pi\colon V \longrightarrow U$ is a branched analytic covering of order r induced by the natural projection mapping $\mathbb{C}^{k+1} \longrightarrow \mathbb{C}^k$ when the variety V is represented by an analytic subvariety of an open neighborhood of the origin in $\mathbb{C}^{k+1}$, then the direct image sheaf $\pi_*({}_V\mathcal{O})$ is a free sheaf of rank r.

In the case of a more general branched analytic covering, the direct image sheaf $\pi_*({}_V\mathcal{O})$ is not necessarily a free analytic sheaf, even locally; however it is always a coherent analytic sheaf. This amounts to the assertion that the kernel sheaf $\mathcal{K}$ in the exact sequence above is locally finitely generated; but actually it is easier to establish that result somewhat more indirectly.

Theorem 19(b). For any branched analytic covering $\pi\colon V \longrightarrow U$, the direct image sheaf $\pi_*({}_V\mathcal{O})$ is a coherent analytic sheaf over U.

Proof. As in the proof of the first part of the theorem, it suffices to consider only a branched analytic covering $\pi\colon V \longrightarrow U$ induced by the natural projection mapping $\mathbb{C}^n \longrightarrow \mathbb{C}^k$, when the variety V is represented by a complex analytic subvariety of an open neighborhood of the origin in $\mathbb{C}^n$. There is no loss of generality in assuming that the coordinates in $\mathbb{C}^n$ are so chosen that the coordinate z_{k+1} generally separates the sheets of the covering. As in the local parametrization theorem, the partial projection $\mathbb{C}^n \longrightarrow \mathbb{C}^{k+1}$ maps the subvariety V onto an analytic subvariety V_o of an open neighborhood of the origin in $\mathbb{C}^{k+1}$, inducing an analytic mapping $\rho: V \longrightarrow V_o$. The further projection $\mathbb{C}^{k+1} \longrightarrow \mathbb{C}^k$ then induces a branched analytic covering $\sigma: V_o \longrightarrow U$, and the original branched analytic covering is the composition $\pi = \sigma\rho$. Introduce the auxiliary subsheaf of rings $\mathcal{S} \subseteq {}_V\mathcal{O}$ on the analytic variety V, defined as having as stalk at any point $p \in V$ the subring $\mathcal{S}_p = {}_k\mathcal{O}_{\pi(p)}[\tilde{z}_{k+1}] \subseteq {}_V\mathcal{O}_p$, where $\tilde{z}_{k+1}$ is the restriction of the coordinate function z_{k+1} to the analytic subvariety V and ${}_k\mathcal{O}_{\pi(p)}$ is as usual identified with its isomorphic image $\pi^*({}_k\mathcal{O}_{\pi(p)}) \subseteq {}_V\mathcal{O}_p$. Letting $d \in {}_k\mathcal{O}_U$ be the discriminant of the polynomial $p_{z_{k+1}}(X) \in {}_k\mathcal{O}_U[X]$ of Theorem 18(a), and noting that this discriminant is not zero since the function z_{k+1} generally separates the sheets of the branched analytic covering $\pi: V \longrightarrow U$, it follows immediately from Theorem 18(b) that $d \cdot {}_V\mathcal{O}_p \subseteq \mathcal{S}_p$ at each point $p \in V$; consequently, in terms of sheaves, $d \cdot {}_V\mathcal{O} \subseteq \mathcal{S} \subseteq {}_V\mathcal{O}$. Note further that clearly $\pi_*(d \cdot {}_V\mathcal{O}) = d \cdot \pi_*({}_V\mathcal{O}) \cong \pi_*({}_V\mathcal{O})$; and therefore $\pi_*({}_V\mathcal{O}) \cong \pi_*(d \cdot {}_V\mathcal{O}) \subseteq \pi_*(\mathcal{S})$. To complete the proof, it suffices to show that $\pi_*(\mathcal{S})$ is a free analytic sheaf; for then $\pi_*({}_V\mathcal{O})$ is exhibited as a subsheaf of a free analytic sheaf, and since

$\pi_*({}_V\mathcal{O})$ is locally a finitely generated analytic sheaf as a consequence of Theorem 19(a), necessarily $\pi_*({}_V\mathcal{O})$ is a coherent analytic sheaf.

Now note first that the direct image sheaf $\rho_*(\mathcal{S}) \cong {}_{V_o}\mathcal{O}$. For selecting any open neighborhood V_{oq} of a point $q \in V_o$, it follows from the definitions of the direct image sheaf and of the sheaf $\mathcal{S}$ that $\Gamma(V_{oq}, \rho_*(\mathcal{S})) \cong \Gamma(\rho^{-1}(V_{oq}), \mathcal{S}) = \Gamma(\rho^{-1}(V_{oq}), {}_U\mathcal{O}[\tilde{z}_{k+1}])$; and since sections of the sheaf ${}_U\mathcal{O}[\tilde{z}_{k+1}]$ are independent of the coordinates $z_{k+2}, \ldots, z_n$, it is further evident that $\Gamma(\rho^{-1}(V_{oq}, {}_U\mathcal{O}[\tilde{z}_{k+1}]) \cong \Gamma(V_{oq}, {}_U\mathcal{O}[\tilde{z}_{k+1}])$. Consequently, upon passing to the germs at the point q, it follows that $\rho_*(\mathcal{S})_q \cong {}_U\mathcal{O}_{\sigma(q)}[\tilde{z}_{k+1}] \cong {}_{V_o}\mathcal{O}_q$. Then $\pi_*(\mathcal{S}) = \sigma_*(\rho_*(\mathcal{S})) \cong \sigma_*({}_{V_o}\mathcal{O})$; and since $\sigma_*({}_{V_o}\mathcal{O})$ is a free analytic sheaf as a consequence of Corollary 1 to Theorem 19, it follows that $\pi_*(\mathcal{S})$ is also a free analytic sheaf, and the proof is thereby concluded.

Although the direct image sheaf $\pi_*({}_V\mathcal{O})$ is not necessarily a few analytic sheaf, it perhaps should be pointed out that the proof of the preceding part of the theorem did provide some slightly more detailed information than merely that the direct image sheaf $\pi_*({}_V\mathcal{O})$ is coherent; for the essential step in the proof was to show that locally the sheaf $\pi_*({}_V\mathcal{O})$ could be imbedded as a subsheaf of a free analytic sheaf. Thus the following assertion is an immediate consequence of the proof of the theorem.

Corollary 2 to Theorem 19. If $\pi: V \longrightarrow U$ is a branched analytic covering of order r, then the direct image sheaf $\pi_*({}_V\mathcal{O})$ is locally an analytic subsheaf of a free analytic sheaf of rank r.

§6. Simple analytic mappings between complex analytic varieties.

(a) The partial projection mappings appearing in the local parametrization theorem have not so far been considered in any detail, although one example of their usefulness was provided in the course of the proof of Theorem 19. Actually these mappings play a very useful role in the study of the local properties of complex analytic varieties. Analytic equivalence is really too strict an equivalence relation to be used from the beginning in attempting an explicit classification of the singularities of complex analytic varieties; it is natural to try to develop a sequence of progressively stricter equivalence relations culminating in analytic equivalence, so that at each stage a more reasonable classification is possible; and the partial projection mappings are of some relevance to this program. The present section will be devoted to a discussion of these partial projection mappings for complex analytic varieties. It is first useful to characterize this class of mappings somewhat more intrinsically.

A <u>simple analytic mapping</u> $\rho\colon V_1 \longrightarrow V$ between two complex analytic varieties V_1 and V is a proper, light, analytic mapping such that there exist analytic subvarieties $A_1 \subset V_1$ and $A \subset V$ for which $V_1 - A_1$ and $V-A$ are dense in V_1 and V respectively and the restriction $\rho\colon V_1 - A_1 \longrightarrow V-A$ is an equivalence of analytic varieties. Note that the image $\rho(V_1)$ is necessarily all of V; for $\rho(V_1)$ contains the dense open subset $V-A \subseteq V$ and $\rho(V_1)$ must be a closed subset of V since the mapping ρ is proper. In most of the applications the varieties V_1 and V are both pure dimensional; of course it is clear that whenever one of the two varieties is pure dimensional, the other variety is also pure dimensional and of the same dimension. That these mappings are at least closely related to the partial projection mappings in the local parametrization theorem is indicated by the following easy observation.

Theorem 20. Suppose that $\rho\colon V_1 \longrightarrow V$ is an analytic mapping between two pure dimensional complex analytic varieties. If ρ is a simple analytic mapping, then there exists an open neighborhood V_q of any point q of V such that the varieties V_q and $V_{1q} = \rho^{-1}(V_q)$ can be represented by branched analytic coverings $\pi_1\colon V_{1q} \longrightarrow U$ and $\pi\colon V_q \longrightarrow U$ of the same order with $\pi_1 = \pi\rho$. Conversely, if there is an open neighborhood V_q of the point q in V such that the varieties V_q and $V_{1q} = \rho^{-1}(V_q)$ can be represented by branched analytic coverings $\pi_1\colon V_{1q} \longrightarrow U$ and $\pi\colon V_q \rightarrow U$ of the same order with $\pi_1 = \pi\rho$, then the restriction $\rho\colon V_{1q} \longrightarrow V_q$ is a simple analytic mapping.

Proof. If ρ is a simple analytic mapping, select any representation $\pi\colon V_q \longrightarrow U$ of an open neighborhood of a point q in V as a branched analytic covering. The analytic mapping $\pi_1 = \pi\rho\colon V_{1q} \longrightarrow U$ is then a light mapping, so by Theorem 16 it is a branched analytic covering provided that V_q is sufficiently small; and clearly the branched analytic coverings $\pi_1\colon V_{1q} \longrightarrow U$ and $\pi\colon V_q \longrightarrow U$ are of the same order. Conversely, if $\pi_1 = \pi\rho$ for some branched analytic coverings $\pi_1\colon V_{1q} \longrightarrow U$ and $\pi\colon V_q \rightarrow U$ of the same order, it is clear that the mapping ρ is both light and proper. Let $B_1 \subset V_{1q}$ and $B \subset V_q$ be the critical loci of these branched analytic coverings, and introduce the analytic subvariety $D = \pi_1(B_1) \cup \pi(B) \subset U$; then $V_{1q} - \pi_1^{-1}(D)$ and $V_q - \pi^{-1}(D)$ are dense open subsets of the varieties V_{1q} and V_q respectively, and are exhibited by π_1 and π as equivalent complex analytic covering manifolds of $U-D$, so that the restriction $\rho\colon V_{1q} - \pi_1^{-1}(D) \longrightarrow V_q - \pi^{-1}(D)$ is an analytic equivalence of varieties. Thus the restriction $\rho\colon V_{1q} \longrightarrow V_q$ is a simple analytic mapping, and the proof is thereby concluded.

<u>Corollary to Theorem 20.</u> If $\rho_1\colon V_1 \longrightarrow V$ and $\rho_2\colon V_2 \rightarrow V_1$ are simple analytic mappings between pure dimensional complex analytic varieties, then the composition $\rho_1\rho_2\colon V_2 \longrightarrow V$ is also a simple analytic mapping.

Proof. If $\rho_1\colon V_1 \longrightarrow V$ is a simple analytic mapping, then locally at least there is a branched analytic covering $\pi\colon V \longrightarrow U$ of order r such that the composition $\pi\rho_1\colon V_1 \longrightarrow U$ is a branched analytic covering of order r . Then, as in the proof of Theorem 20, the composition $\pi\rho_1\rho_2\colon V_2 \longrightarrow U$ is also a branched analytic covering of order r , so that $\rho_1\rho_2$ is a simple analytic mapping as desired.

It should be observed that a simple analytic mapping $\rho\colon V_1 \longrightarrow V$ is superficially almost the same thing as a branched analytic covering of order 1, except of course that the range space V is not necessarily just an open subset of the complex number space. However, there is really quite a considerable difference between these two concepts; for instance, a branched analytic covering of order 1 is necessarily a complex analytic equivalence, whereas the partial projection mappings in the local parametrization theorem are examples of simple analytic mappings which are not necessarily analytic equivalences. Indeed, a simple analytic mapping ρ is not even necessarily a one-to-one mapping.

To see how this can happen, select any point $q \in V$, and as in Theorem 20 choose an open neighborhood V_q of q in V such that there are branched analytic coverings $\pi_1\colon V_{1q} \longrightarrow U$ and $\pi\colon V_q \longrightarrow U$ of the same order with $V_{1q} = \rho^{-1}(V_q)$ and $\pi_1 = \pi\rho$. If the neighborhood V_q is chosen sufficiently small then $q = \pi^{-1}(\pi(q))$. Let $\pi_1^{-1}(\pi(q)) = \rho^{-1}(q) = \{p_1,\ldots,p_s\} \subset V_{1q}$; and applying the localization lemma, it can be assumed that $\pi_1^{-1}(U) = \rho^{-1}(V_q)$ consists of s connected components $V_{1q}^{(1)},\ldots,V_{1q}^{(s)}$ such that $p_i \in V_{1q}^{(i)}$ and the restriction $\pi_1\colon V_{1q}^{(i)} \longrightarrow U$ is also a branched analytic covering. If $B_1 \subset V_{1q}$ and $B \subset V_q$ are the critical loci of these branched analytic coverings, then $D = \pi_1(B_1) \cup \pi(B)$ is an analytic subvariety of U; and over the complement $U-D$ both π_1 and π are analytic covering projections in the usual sense. Recall from the local parametrization theorem that if the neighborhood V_q is connected and sufficiently small, then the point set closures of the connected components of $\pi_1^{-1}(U-D) \subseteq V_{1q}$ and of $\pi^{-1}(U-D) \subseteq V_q$ are analytic subvarieties of V_{1q} and V_q respectively, representing the irreducible components of the germs of these varieties at the various points $p_1,\ldots,p_s,q$. Now if $s > 1$, it is clear that $\pi_1^{-1}(U-D)$ has at least s connected components corresponding to the s components of $\pi_1^{-1}(U)$; and consequently $\pi^{-1}(U-D)$ also has at least s connected components, so that V is necessarily reducible at the point q . Moreover the component $V_{1q}^{(i)}$ of the variety V_{1q} containing the point p_i is evidently mapped analytically by ρ to an analytic subvariety $V_q^{(i)} \subset V_q$, such that the restriction $\rho\colon V_{1q}^{(i)} \longrightarrow V_q^{(i)}$ is itself a simple analytic mapping; and the varieties $V_q^{(i)}$ clearly contain no common irreducible components, so that the decomposition $V_q = \cup_i V_q^{(i)}$ is some grouping of the irreducible components of V_q at the point q . Thus geometrically, if $\rho^{-1}(q)$ contains more than one point, then V is reducible at q and the separate components of V_1 at the distinct points of $\rho^{-1}(q)$ are mapped to separate component varieties of V .

Note that if $\rho\colon V_1 \longrightarrow V$ is a simple analytic mapping, and if $V_q \subseteq V$ is an open neighborhood of a point $q \in V$, then the restriction $\rho\colon \rho^{-1}(V_q) \longrightarrow V_q$ is also a simple analytic mapping. Thus it is possible to introduce the notion of the germ of a simple analytic mapping $\rho\colon V_1 \longrightarrow V$ over the germ V of a complex analytic variety, observing that V_1 may necessarily consist of a finite number of germs of complex analytic varieties.

(b) If $\rho\colon V_1 \longrightarrow V$ is a simple analytic mapping, then for each point $p_1 \in V_1$ there is a natural ring homomorphism $\rho_1^*\colon {}_V\mathcal{O}_{\rho(p_1)} \longrightarrow {}_{V_1}\mathcal{O}_{p_1}$. It is quite easy to see that this homomorphism is an isomorphism into the ring ${}_{V_1}\mathcal{O}_{p_1}$ precisely when $\rho^{-1}(\rho(p_1)) = p_1$. For if $\rho^{-1}(\rho(p_1))$ contains points other than p_1, the observations made in the preceding paragraphs show that the image under ρ of the germ of the variety V_1 at the point p_1 is the germ of a proper analytic subvariety $V' \subset V$ of the germ of the variety V at the point $\rho(p_1)$; and selecting a non trivial germ $f \in {}_V\mathcal{O}_{\rho(p_1)}$ which vanishes identically on the subvariety $V' \subset V$, it is clear that $\rho_1^*(f) = f \circ \rho = 0$ in ${}_{V_1}\mathcal{O}_{p_1}$. On the other hand, if $p_1 = \rho^{-1}(\rho(p_1))$ then the image under ρ of the germ of the variety V_1 at the point p_1 is the entire germ of the variety V at the point $\rho(p_1)$, and consequently ρ_1^* is an isomorphism into ${}_{V_1}\mathcal{O}_{p_1}$. Now selecting a point $q \in V$ and letting $\rho^{-1}(q) = \{p_1,\ldots,p_s\} \subset V_1$, the homomorphisms $\rho_i^*\colon {}_V\mathcal{O}_q \longrightarrow {}_{V_1}\mathcal{O}_{p_i}$ can be considered as determining a single ring homomorphism

$$\rho^*\colon {}_V\mathcal{O}_q \longrightarrow {}_{V_1}\mathcal{O}_{p_1} \oplus \cdots \oplus {}_{V_1}\mathcal{O}_{p_s}$$

into the direct sum of the various rings ${}_{V_1}\mathcal{O}_{p_i}$; and this homomorphism is clearly an isomorphism into that direct sum ring.

Note that when $s > 1$ the image $\rho^*({}_V\mathcal{O}_q)$ is necessarily a proper subring of ${}_{V_1}\mathcal{O}_{p_1} \oplus \cdots \oplus {}_{V_1}\mathcal{O}_{p_s}$; for if $f \in {}_V\mathcal{O}_q$ and $\rho^*(f) = (f_1,\ldots,f_s) \in {}_{V_1}\mathcal{O}_{p_1} \oplus \cdots \oplus {}_{V_1}\mathcal{O}_{p_s}$, then

$f_i(p_i) = f(\rho_i(p_i)) = f(q)$ is independent of i. Even when $s = 1$, the image $\rho^*({}_V\mathcal{O}_q)$ may be a proper subring; indeed, in this case it is easy to see that $\rho^*({}_V\mathcal{O}_q) = {}_{V_1}\mathcal{O}_p$ if and only if the mapping ρ is an analytic equivalence between the germ of the variety V_1 at p and the germ of the variety V at $q = \rho(p)$. In any case, though, the image $\rho^*({}_V\mathcal{O}_q)$ cannot really be too small, in a sense that can be made precise through the following discussion.

A <u>relative denominator</u> for the simple analytic mapping $\rho: V_1 \longrightarrow V$ at a point $q \in V$ is an element $d \in {}_V\mathcal{O}_q$ such that

$$\rho^*(d)\cdot({}_{V_1}\mathcal{O}_{p_1} \oplus \cdots \oplus {}_{V_1}\mathcal{O}_{p_s}) \subseteq \rho^*({}_V\mathcal{O}_q) ,$$

where as before $\rho^{-1}(q) = \{p_1,\ldots,p_s\} \in V_1$; that is to say, a relative denominator is an element $d \in {}_V\mathcal{O}_q$ with the property that for any germs $f_i \in {}_{V_1}\mathcal{O}_{p_i}$, $i = 1,\ldots,s$, there exists a germ $f \in {}_V\mathcal{O}_q$ such that $\rho_i^*(d)\cdot f_i = \rho_i^*(f) \in {}_{V_1}\mathcal{O}_{p_i}$, for $i = 1,\ldots,s$. Note that the zero element of ${}_V\mathcal{O}_q$ is a relative denominator, although of course in a rather trivial sense; but at least the set of relative denominators is not empty. It is clear that the set of all relative denominators for the mapping ρ form an ideal in the local ring ${}_V\mathcal{O}_q$; this ideal will be called <u>the ideal of relative denominators</u> for the simple analytic mapping ρ at the point q , and will be denoted by $\mathcal{J}(\rho)_q$. Note further that when $\mathcal{J}(\rho)_q = {}_V\mathcal{O}_q$, so that in particular the constant function 1 is a relative denominator, then $\rho^*({}_V\mathcal{O}_q) = {}_{V_1}\mathcal{O}_{p_1} \oplus \cdots \oplus {}_{V_1}\mathcal{O}_{p_s}$; but this means that $\rho^{-1}(q) = p$ is a single point of V_1 , and that the mapping $\rho: V_1 \longrightarrow V$ is an analytic equivalence between the germ of the variety V_1 at p and the germ of the variety V at q, as noted above. As an extension of the definition, an element $d \in {}_V\mathcal{O}_q$ will be called a <u>universal denominator</u> if it is a relative denominator at the point q for any germ of a simple analytic mapping $\rho: V_1 \longrightarrow V$ at the point $q \in V$. Again the zero element of ${}_V\mathcal{O}_q$ is a universal denominator, and the set of all universal denominators form an ideal in the local ring ${}_V\mathcal{O}_q$; this ideal will be called <u>the ideal of universal denominators</u> for the variety V at the point q, and will be denoted by $\mathcal{J}_q$. That there are in fact non trivial universal denominators is a consequence of the following result.

<u>Theorem 21.</u> There exists a holomorphic function d in an open neighborhood of any point of a pure dimensional complex analytic variety, such that d is a universal denominator but not a zero divisor at each point of that neighborhood.

Proof. Choose an open neighborhood W of the given point such that W can be represented by a branched analytic covering $\pi: W \longrightarrow U$ of order r and that there exists a holomorphic function $g \in {}_W\mathcal{O}_W$ which separates the sheets of this branched analytic covering. The polynomial $p_g(X) \in {}_k\mathcal{O}_U[X]$ of Theorem 18(a) then has a discriminant $d_g \in {}_k\mathcal{O}_U$ which is not identically zero; and the function $d = \pi^*(d_g) \in {}_W\mathcal{O}_W$ is holomorphic on W and is not a zero divisor at any point of W. The proof will be completed by showing that this function d is a universal denominator at each point of W.

Consider therefore a simple analytic mapping $\rho: V_1 \longrightarrow W_q$ over an open neighborhood W_q of some point q in W. By the localization lemma the neighborhood W_q can be so chosen that the restriction $\pi: W_q \longrightarrow \pi(W_q) \subseteq U$ is also a branched analytic covering, although perhaps of order less than r. The restriction of the function g to the neighborhood W_q still separates sheets, and the polynomial $p_g'(X) \in {}_k\mathcal{O}_{\pi(W_q)}[X]$ associated to this restricted branched analytic covering as in Theorem 18(a) is evidently a factor of the full polynomial $p_g(X)$; hence the discriminant $d' \in {}_k\mathcal{O}_{\pi(W_q)}$ of the polynomial $p_g'(X)$ is a factor of the discriminant $d \in {}_k\mathcal{O}_{\pi(W_q)}$, so that $d = d'\cdot d''$ for some holomorphic function $d'' \in {}_k\mathcal{O}_{\pi(W_q)}$. Note that $\pi\rho: V_1 \longrightarrow \pi(W_q)$ is also a branched analytic covering, that the induced function $\rho^*(g) \in {}_{V_1}\mathcal{O}_{V_1}$ also separates sheets, and that the polynomial $p_{\rho^*(g)}(X) \in {}_k\mathcal{O}_{\pi(W_q)}[X]$ associated to this function as in Theorem 18(a) also has the discriminant d'. If $\rho^{-1}(q) = \{p_1,\ldots,p_r\} \subset V_1$, it can be assumed that the neighborhood W_q is chosen such that $\rho^{-1}(W_q)$ consists of s components $V_{p_1},\ldots,V_{p_s}$ with $p_i \in V_{p_i}$; and given any elements $f_i \in {}_{V_1}\mathcal{O}_{p_i}$, after shrinking the neighborhood W_q if necessary these germs will be represented by holomorphic functions f_i on the various components V_{p_i}. These functions together form a single holomorphic function f on $\rho^{-1}(W_q)$; and it follows from Theorem 18(b) that $d'\cdot f$ can be expressed as a polynomial in $\rho^*(g)$ with coefficients in ${}_k\mathcal{O}_{\pi(W_q)}$, hence that $d'\cdot f \in \rho^*({}_W\mathcal{O}_q)$. Therefore

$d \cdot f = d'' \cdot d' f \in \rho^*({}_W\mathcal{O}_q)$, so that d is a relative denominator for the simple analytic mapping ρ at the point q , and the proof is thereby concluded.

The terms relative denominator and universal denominator are suggested by the interpretation of these concepts by means of meromorphic functions. Recall that at a point q of a complex analytic variety V at which that variety is irreducible, the local ring ${}_V\mathcal{O}_q$ is an integral domain; and the elements of the field of quotients ${}_V\mathcal{M}_q$ of this integral domain are defined to be the germs of meromorphic functions on the variety V at the point q . At a point q at which the variety is reducible, the local ring ${}_V\mathcal{O}_q$ is not an integral domain; but it is still possible to introduce the <u>total quotient ring</u> ${}_V\mathcal{M}_q$ of the local ring ${}_V\mathcal{O}_q$, and the elements of ${}_V\mathcal{M}_q$ are defined to be the <u>germs of meromorphic functions</u> on the variety V at the point q . To recall this construction, in case it should not be familiar, introduce the ideal ${}_V\mathcal{N}_q \subset {}_V\mathcal{O}_q$ consisting of all zero divisors in the local ring ${}_V\mathcal{O}_q$; the total quotient ring ${}_V\mathcal{M}_q$ is the ring of all formal quotients f/g where $f \in {}_V\mathcal{O}_q$ and $g \in {}_V\mathcal{O}_q - {}_V\mathcal{N}_q$, with the usual definitions of equivalence and of the ring operations. The ring ${}_V\mathcal{O}_q$ is naturally imbedded in ${}_V\mathcal{M}_q$ as the subring consisting of all formal quotients $f/1$. The units of the ring ${}_V\mathcal{M}_q$ consist of those formal quotients f/g for which $f \notin {}_V\mathcal{N}_q$; hence ${}_V\mathcal{M}_q$ is a field precisely when ${}_V\mathcal{N}_q = 0$, or equivalently, precisely when V is irreducible at q .

Now for any simple analytic mapping $\rho: V_1 \longrightarrow V$ and any points $p_1 \in V_1$ and $q = \rho(p_1) \in V$, consider the ring homomorphism $\rho_1^*: {}_V\mathcal{O}_q \longrightarrow {}_{V_1}\mathcal{O}_{p_1}$. Note that if $g \in {}_V\mathcal{O}_q$ is not a zero divisor, then $\rho_1^*(g) \in {}_{V_1}\mathcal{O}_{p_1}$ is not a zero divisor either; for $\rho_1^*(g)$ is a zero divisor only when $\rho_1^*(g)$ vanishes on one of the irreducible components of the analytic variety V_1 at the point p_1, and that can only happen when g vanishes on one of the irreducible components of the analytic variety V at the point q, hence when g is a zero divisor in ${}_V\mathcal{O}_q$. Thus the homomorphism $\rho_1^*: {}_V\mathcal{O}_q \longrightarrow {}_{V_1}\mathcal{O}_{p_1}$ induces a ring homomorphism $\rho^*: {}_V\mathcal{M}_q \longrightarrow {}_{V_1}\mathcal{M}_{p_1}$. If $\rho^{-1}(q) = \{p_1, \ldots, p_s\} \subset V$, then the various homomorphisms $\rho_i^*: {}_V\mathcal{M}_q \longrightarrow {}_{V_1}\mathcal{M}_{p_i}$ can be considered as determining a single ring homomorphism

$$\rho^*: {}_V\mathcal{M}_q \longrightarrow {}_{V_1}\mathcal{M}_{p_1} \oplus \cdots \oplus {}_{V_1}\mathcal{M}_{p_s}$$

into the direct sum of the various rings ${}_{V_1}\mathcal{M}_{p_i}$. As in the case of holomorphic functions, it is clear that this mapping is always an isomorphism into the direct sum ring; but for the case of meromorphic functions, the existence of a relative denominator which is not a zero divisor implies that this isomorphism is onto the full direct sum ring, hence that

$${}_V\mathcal{M}_q \cong {}_{V_1}\mathcal{M}_{p_1} \oplus \cdots \oplus {}_{V_1}\mathcal{M}_{p_s}$$

under the isomorphism ρ^*. To see this, select any relative denominator $d \in \mathcal{S}(\rho)_q \subseteq {}_V\mathcal{O}_q$ which is not a zero divisor in the ring

${}_V\mathcal{O}_q$, and consider any meromorphic functions $f_i/g_i \in {}_{V_1}\mathcal{M}_{p_i}$, $i = 1,\dots,s$. Since d is a relative denominator, there exist holomorphic functions f and g in ${}_V\mathcal{O}_q$ such that $\rho_i^*(f) = \rho_i^*(d)\cdot f_i$ and $\rho_i^*(g) = \rho_i^*(d)\cdot g_i$ for $i = 1,\dots,s$; evidently g is not a zero divisor in ${}_V\mathcal{O}_q$, and the meromorphic function $f/g \in {}_V\mathcal{M}_q$ has the property that $\rho_i^*(f/g) = f_i/g_i$ for $i = 1,\dots,s$, as desired. In particular, given any holomorphic functions $f_i \in {}_{V_1}\mathcal{O}_{p_i} \subset {}_{V_1}\mathcal{M}_{p_i}$, there will at least exist a meromorphic function $f/g \in {}_V\mathcal{M}_q$ such that $\rho_i^*(f/g) = f_i$ for $i = 1,\dots,s$; and indeed, it can always be assumed that the denominator g of this meromorphic function is any assigned relative denominator $g \in \mathcal{J}(\rho)_q \subseteq {}_V\mathcal{O}_q$ which is not a zero divisor in the ring ${}_V\mathcal{O}_q$.

(c) As in the case of branched analytic coverings, so also in the case of simple analytic mappings is it more convenient to consider direct images rather than inverse images. If $\rho\colon V_1 \longrightarrow V$ is a simple analytic mapping, the direct image sheaf $\rho_*({}_{V_1}\mathcal{O})$ is evidently a well defined sheaf of modules over the sheaf of rings ${}_V\mathcal{O}$ on the analytic variety V , thus an analytic sheaf over the variety V . In a similar manner, introducing the sheaves ${}_V\mathcal{M}$ and ${}_{V_1}\mathcal{M}$ of germs of meromorphic functions on the analytic varieties V and V_1 respectively, which are clearly analytic sheaves on their respective varieties, the direct image sheaf $\rho_*({}_{V_1}\mathcal{M})$ is also an analytic sheaf over the variety V . Recalling the definition of the direct image sheaf, if $\rho^{-1}(q) = \{p_1,\ldots,p_s\} \subset V_1$ for a point $q \in V$, it is clear that $\rho_*({}_{V_1}\mathcal{O})_q \cong {}_{V_1}\mathcal{O}_{p_1} \oplus\ldots\oplus {}_{V_1}\mathcal{O}_{p_s}$ and that $\rho_*({}_{V_1}\mathcal{M})_q \cong {}_{V_1}\mathcal{M}_{p_1} \oplus\ldots\oplus {}_{V_1}\mathcal{M}_{p_s}$. Since ${}_V\mathcal{M}_q \cong {}_{V_1}\mathcal{M}_{p_1} \oplus\ldots\oplus {}_{V_1}\mathcal{M}_{p_s}$ under the inverse isomorphism ρ^*, as noted earlier, it follows that it is possible to identify the direct image sheaf $\rho_*({}_{V_1}\mathcal{M})$ with the sheaf ${}_V\mathcal{M}$ over the variety V; that is to say, the direct image sheaf $\rho_*({}_{V_1}\mathcal{M})$ is canonically isomorphic to the sheaf ${}_V\mathcal{M}$ itself. Since ${}_{V_1}\mathcal{O} \subset {}_{V_1}\mathcal{M}$, it follows that under this isomorphism the direct image sheaf $\rho_*({}_{V_1}\mathcal{O})$ is canonically isomorphic to an analytic subsheaf of the sheaf ${}_V\mathcal{M}$ of germs of meromorphic functions on V. To be quite explicit, an element $F \in \rho_*({}_{V_1}\mathcal{O})_q$ is described in the usual way by a set of germs $f_i \in {}_{V_1}\mathcal{O}_{p_i}$ for $i=1,\ldots,s$; there is, however, a unique meromorphic function $f/g \in {}_V\mathcal{M}_q$ such that $\rho_i^*(f/g) = f_i$ for $i=1,\ldots,s$, and the element F will be identified with this meromorphic function. Thus the elements of the stalk $\rho_*({}_{V_1}\mathcal{O})_q$ will be identified with the set of those meromorphic functions $f/g \in {}_V\mathcal{M}_q$ such that $\rho_i^*(f/g) \in {}_{V_1}\mathcal{O}_{p_i}$ for $i=1,\ldots,s$; note that the denominator g can be taken to be any preassigned relative denominator for the simple analytic mapping ρ, provided only that g is not a zero divisor.

Theorem 22. Suppose that $\rho\colon V_1 \longrightarrow V$ is a simple analytic mapping between two pure dimensional complex analytic varieties.

(a) The direct image sheaf $\rho_*({}_{V_1}\mathcal{O})$ is a coherent analytic sheaf over the variety V, locally isomorphic to a sheaf of ideals in the structure sheaf ${}_V\mathcal{O}$.

(b) The sheaf $\mathcal{J}(\rho)$ of ideals of relative denominators is also a coherent analytic sheaf over the variety V .

Proof. (a) Since the theorem is local in character, there is no loss of generality in considering merely an arbitrarily small open neighborhood of some point of the variety V . Thus as a consequence of Theorem 21 it can be assumed that there is a holomorphic function d on V which is a relative denominator at each point but is nowhere a zero divisor. The direct image sheaf $\rho_*({}_{V_1}\mathcal{O})$ can be identified with an analytic subsheaf of the sheaf ${}_V\mathcal{M}$ of germs of meromorphic functions over V, and indeed, the elements of $\rho_*({}_{V_1}\mathcal{O})_q \subset {}_V\mathcal{M}_q$ can all be taken to have the common denominator $d \in {}_V\mathcal{O}_q$ at any point $q \in V$; thus $\rho_*({}_{V_1}\mathcal{O}) \cong d\cdot\rho_*({}_{V_1}\mathcal{O}) \subseteq {}_V\mathcal{O}$, so that $\rho_*({}_{V_1}\mathcal{O})$ is isomorphic to a sheaf of ideals over the variety V . To demonstrate the coherence of the sheaf $\rho_*({}_{V_1}\mathcal{O})$ it then suffices merely to show that that sheaf is locally finitely generated. As consequences of Theorem 20 and Theorem 19(a), it can further be assumed that there are branched analytic coverings $\pi\colon V \longrightarrow U$ and $\pi_1\colon V_1 \longrightarrow U$ for which $\pi_1 = \pi\rho$ and $\pi_{1*}({}_{V_1}\mathcal{O})$ is a finitely generated analytic sheaf over U . Choosing functions $h_\nu \in \Gamma(V_1, {}_{V_1}\mathcal{O}) \cong \Gamma(U, \pi_{1*}({}_{V_1}\mathcal{O}))$ which represent generators of the sheaf $\pi_{1*}({}_{V_1}\mathcal{O})$ as an analytic sheaf over U , the proof of this portion of the theorem will be completed by showing that these functions $h_\nu \in \Gamma(V_1, {}_{V_1}\mathcal{O}) \cong \Gamma(V, \rho_*({}_{V_1}\mathcal{O}))$ represent generators of the sheaf $\rho_*({}_{V_1}\mathcal{O})$ as an analytic sheaf over V. For any point $q \in V$ let $\rho^{-1}(q) = \{p_1, \ldots, p_s\} \subset V$, and $\pi_1^{-1}\pi(q) = \{p_1, \ldots, p_s, p_{s+1}, \ldots, p_t\} \subset V_1$. Now an element $F \in \rho_*({}_{V_1}\mathcal{O})_q$ is represented by germs $f_i \in {}_{V_1}\mathcal{O}_{p_i}$ for $i = 1, \ldots, s$; and these germs together with the zero germs $0 \in {}_{V_1}\mathcal{O}_{p_i}$ for $i = s+1, \ldots, t$ represent in turn an element $\hat{F} \in \pi_{1*}({}_{V_1}\mathcal{O})_{\pi(q)}$. Since the functions h_ν represent generators of the analytic sheaf $\pi_{1*}({}_{V_1}\mathcal{O})$, there exist germs $g_\nu \in {}_U\mathcal{O}_{\pi(q)}$ such that $\hat{F} = \Sigma_\nu g_\nu h_\nu$; but then $f_i = \Sigma_\nu \pi_1^*(g_\nu)h_\nu \in {}_{V_1}\mathcal{O}_{p_i}$ for $i = 1, \ldots, s$, so that $F = \Sigma_\nu \pi^*(g_\nu)h_\nu \in \rho_*({}_{V_1}\mathcal{O})_q$, and the desired result is thereby demonstrated.

(b) Again as a consequence of Theorem 21 it can be assumed that there is a holomorphic function d on V which is a relative denominator at each point but is nowhere a zero divisor, since the second part of the theorem is also local in character. The direct image sheaf $\rho_*({}_{V_1}\mathcal{O})$ can be identified with an analytic subsheaf of the sheaf ${}_V\mathcal{M}$ of germs of meromorphic functions on V, and $d \cdot \rho_*({}_{V_1}\mathcal{O}) \cong \rho_*({}_{V_1}\mathcal{O})$ is a sheaf of ideals on V. Note that a germ $g \in {}_V\mathcal{O}_q$ is a relative denominator for the simple analytic mapping ρ at a point $q \in V$ precisely when $dg \cdot \rho_*({}_{V_1}\mathcal{O})_q \subset d \cdot {}_V\mathcal{O}_q$. (It is evident from the definition that a germ $g \in {}_V\mathcal{O}_q$ is a relative denominator precisely when $g \cdot \rho_*({}_{V_1}\mathcal{O})_q \subseteq {}_V\mathcal{O}_q$, identifying $\rho_*({}_{V_1}\mathcal{O})_q$ with a submodule of ${}_V\mathcal{M}_q$. Thus if g is a relative denominator, then $dg\cdot\rho_*({}_{V_1}\mathcal{O})_q \subseteq d\cdot{}_V\mathcal{O}_q$; and conversely if $dg\cdot\rho_*({}_{V_1}\mathcal{O})_q \subseteq d\cdot{}_V\mathcal{O}_q$, then since d is not a zero divisor necessarily $g\cdot\rho_*({}_{V_1}\mathcal{O})_q \subseteq {}_V\mathcal{O}_q$, and hence g is a relative denominator.) Thus

$$\mathcal{J}(\rho)_q = \{g \in {}_V\mathcal{O}_q \mid g\cdot d\cdot\rho_*({}_{V_1}\mathcal{O})_q \subseteq d\cdot{}_V\mathcal{O}_q\} ;$$

and since both $d\cdot\rho_*({}_{V_1}\mathcal{O})$ and $d\cdot{}_V\mathcal{O}$ are coherent sheaves of ideals over V , it follows that $\mathcal{J}(\rho)$ is also a coherent sheaf of ideals over V . (Although this last step is a standard argument, an additional few words might prove helpful to the beginner. The sheaf $d\cdot\rho_*({}_{V_1}\mathcal{O})$ is a coherent sheaf of ideals by the first part of the theorem; so select finitely many functions $h_\nu \in \Gamma(V, {}_V\mathcal{O})$ which generate that sheaf of ideals at each point of V , restricting V if necessary. The residue class sheaf ${}_V\mathcal{O}/d\cdot{}_V\mathcal{O}$ is of course a coherent analytic sheaf also. The above formula shows that $\mathcal{J}(\rho)$ is the kernel of the analytic sheaf homomorphism

$$\varphi:\ {}_V\mathcal{O} \longrightarrow \oplus_\nu({}_V\mathcal{O}/d\cdot{}_V\mathcal{O})$$

which associates to an element $f \in {}_V\mathcal{O}_q$ the residue classes of the elements $h_\nu f$ in ${}_V\mathcal{O}/d\cdot{}_V\mathcal{O}$, hence $\mathcal{J}(\rho)$ is a coherent sheaf of ideals as desired.)

For a simple analytic mapping $\rho: V_1 \longrightarrow V$, the ideal $\mathcal{J}(\rho)_q \subseteq {}_V\mathcal{O}_q$ at a point $q \in V$ determines the germ of an analytic subvariety $\operatorname{loc}\mathcal{J}(\rho)_q$ of the variety V at that point. Actually, since $\mathcal{J}(\rho)$ is a coherent sheaf of ideals, there is a well defined analytic subvariety $\operatorname{loc}\mathcal{J}(\rho) \subset V$ such that at any point $q \in V$

the germ of the subvariety loc $\mathcal{D}(\rho)$ is just loc $\mathcal{D}(\rho)_q$; for in an open neighborhood of any point of the variety V there are finitely many holomorphic functions which generate the ideal $\mathcal{D}(\rho)_q \subseteq {}_V\mathcal{O}_q$ at each point q of that neighborhood, hence the set of common zeros of those functions is the subvariety loc $\mathcal{D}(\rho)$ in that neighborhood. Recall that at a point $q \in V$ - loc $\mathcal{D}(\rho)$, hence at a point $q \in V$ for which $\mathcal{D}(\rho)_q = {}_V\mathcal{O}_q$, the inverse image $\rho^{-1}(q)$ is a single point of V_1 and the analytic mapping ρ is an analytic equivalence between the germ of the variety V_1 at $\rho^{-1}(q)$ and the germ of the variety V at the point q ; while at a point $q \in$ loc $\mathcal{D}(\rho)$, the mapping ρ cannot be an analytic equivalence. This can be summarized as follows.

<u>Corollary 1 to Theorem 22.</u> For a simple analytic mapping $\rho: V_1 \longrightarrow V$ between two pure dimensional complex analytic varieties, the set of points of V at which the mapping ρ is not an analytic equivalence is precisely the analytic subvariety loc $\mathcal{D}(\rho) \subset V$.

At a regular point $q \in V$ a simple analytic mapping $\rho: V_1 \longrightarrow V$ is of course precisely the same thing as a branched analytic covering of order 1, hence is an analytic equivalence; so as a consequence of the preceding corollary $\mathcal{R}(V) \subseteq V$ - loc $\mathcal{D}(\rho)$. This can be restated more conveniently as follows.

<u>Corollary 2 to Theorem 22.</u> For a simple analytic mapping $\rho: V_1 \longrightarrow V$ between two pure dimensional complex analytic varieties,

loc $\mathcal{D}(\rho) \subseteq \mathcal{S}(V)$, where as usual $\mathcal{S}(V)$ is the singular locus of the variety V .

Applying the Hilbert zero theorem, it follows from Corollary 2 that at any point $q \in V$,

$$\sqrt{\mathcal{D}(\rho)_q} = \text{id loc}\, \mathcal{D}(\rho)_q \supseteq \text{id}\, \mathcal{S}(V)_q \ .$$

Consequently any germ $f \in \text{id}\, \mathcal{S}(V)_q \subseteq {}_V\mathcal{O}_q$ must be contained in the radical of the ideal $\mathcal{D}(\rho)_q \subseteq {}_V\mathcal{O}_q$, so that a power of the germ f is a relative denominator for the simple analytic mapping ρ at the point $q \in V$. This observation is also worth restating explicitly as follows.

Corollary 3 to Theorem 22. For a simple analytic mapping $\rho: V_1 \longrightarrow V$ between two pure dimensional complex analytic varieties, some power f^ν of any germ $f \in {}_V\mathcal{O}_q$ which vanishes on the singular locus $\mathcal{S}(V) \subset V$ at the point q is a relative denominator for ρ at the point q .

(d) Under a simple analytic mapping $\rho: V_1 \longrightarrow V$, the direct image $\rho_*({}_{V_1}\mathcal{O})_q$ at a point $q \in V$ has been canonically identified with a submodule of the ${}_V\mathcal{O}_q$ -module ${}_V\mathcal{M}_q$ of germs of meromorphic functions on the variety V at the point q . This direct image module completely characterizes the germ of the simple analytic mapping at the point q , as can be seen readily from the following theorem.

<u>Theorem 23.</u> If $\rho_1: V_1 \longrightarrow V$ and $\rho_2: V_2 \longrightarrow V$ are germs of simple analytic mappings over the pure dimensional complex analytic space V at a point $q \in V$, and if $\rho_{2*}({}_{V_2}\mathcal{O})_q \subseteq \rho_{1*}({}_{V_1}\mathcal{O})_q$, then there is a germ of a complex analytic mapping $F: V_1 \longrightarrow V_2$ such that $\rho_2 F = \rho_1$.

Proof. Restricting the variety V to a sufficiently small open neighborhood of the point q, it can be assumed that $\rho_1: V_1 \longrightarrow V$ and $\rho_2: V_2 \longrightarrow V$ are simple analytic mappings, and that the variety V_2 is represented by a complex analytic subvariety V_2 of an open subset of $\mathbf{C}^{n_2}$. If $w_1, \ldots, w_{n_2}$ are the coordinate functions in $\mathbf{C}^{n_2}$, the restrictions $w_j | V_2$ are holomorphic functions on V_2 and $\rho_{2*}(w_j | V_2)_q \in \rho_{2*}({}_{V_2}\mathcal{O})_q \subseteq \rho_{1*}({}_{V_1}\mathcal{O})_q$; hence there are holomorphic functions f_j on the variety V_1, after restricting V to a still smaller neighborhood of the point q if necessary, such that $\rho_{2*}(w_j | V_2)_q = \rho_{1*}(f_j)_q$. These functions $f_1, \ldots, f_{n_2}$ define a complex analytic mapping F from the analytic variety V_1 into $\mathbf{C}^{n_2}$; and the proof will be concluded by showing that this is the desired mapping. Recall that there is a complex analytic subvariety $A \subset V$ such that the restrictions $\rho_1: V_1 - A_1 \longrightarrow V - A$ and $\rho_2: V_2 - A_2 \longrightarrow V - A$ are complex analytic equivalences between dense open subsets of the varieties V, V_1, V_2, where $A_1 = \rho_1^{-1}(A)$ and $A_2 = \rho_2^{-1}(A)$. By construction, the restriction of the mapping F to the subset $V_1 - A_1$ coincides with the analytic homeomorphism $\rho_2^{-1}\rho_1: V_1 - A_1 \longrightarrow V_2 - A_2$; hence by continuity, F itself is a complex analytic mapping $F: V_1 \longrightarrow V_2$

such that $\rho_2 F = \rho_1$, and that completes the proof.

An immediate consequence of this theorem is then the desired result.

Corollary to Theorem 23. If $\rho_1 \colon V_1 \longrightarrow V$ and $\rho_2 \colon V_2 \rightarrow V$ are germs of simple analytic mappings over the pure dimensional complex analytic variety V at a point $q \in V$, and if $\rho_{1*}({}_{V_1}\mathcal{O})_q = \rho_{2*}({}_{V_2}\mathcal{O})_q$, then the germ of V_1 at $\rho_1^{-1}(q)$ is analytically equivalent to the germ of V_2 at $\rho_2^{-1}(q)$, under the germ of an analytic mapping $F \colon V_1 \longrightarrow V_2$ such that $\rho_2 F = \rho_1$.

Since these direct image modules $\rho_*({}_{V_1}\mathcal{O})_q$ do characterize the germs of simple analytic mappings $\rho \colon V_1 \longrightarrow V$, for the classification of all the simple analytic mappings over V at a point q it suffices to determine which submodules of ${}_V\mathcal{M}_q$ can arise as direct images $\rho_*({}_{V_1}\mathcal{O})_q$. It is evident that the elements of $\rho_*({}_{V_1}\mathcal{O})_q \subset {}_V\mathcal{M}_q$ are bounded meromorphic functions in a neighborhood of q , at least at those points at which their values are well defined; and it will be demonstrated that every bounded meromorphic function in ${}_V\mathcal{M}_q$ is in the direct image $\rho_*({}_{V_1}\mathcal{O})_q$ for some simple analytic mapping $\rho \colon V_1 \longrightarrow V$ in a neighborhood of q . It is first necessary to establish a few further properties of meromorphic functions on an analytic variety.

On a complex analytic manifold, all bounded meromorphic functions are of course holomorphic; indeed, rather more generally, it follows from the Riemann removable singularities theorem that all bounded holomorphic functions on the complement of a proper analytic subvariety of a connected analytic manifold extend uniquely to holomorphic functions on the entire manifold. This is not the case for an arbitrary complex analytic variety, though; but at least the following does hold.

Removable Singularities Lemma. (a). Let W be a proper analytic subvariety of a pure dimensional complex analytic variety V , and let f be a bounded holomorphic function in the intersection of an open neighborhood of the point q in V with the complement $V-W$. Then f represents the germ of a meromorphic function $f \in {}_V\mathcal{M}_q$; and there is a monic polynomial $p_f(X) \in {}_V\mathcal{O}_q[X]$ such that $p_f(f) = 0$ in ${}_V\mathcal{M}_q$.

Proof. Since the desired result is of a local character, there is no loss of generality in supposing that the variety V is represented as a branched analytic covering $\pi\colon V \longrightarrow U$. The usual Riemann removable singularities theorem shows that the function f extends to a holomorphic function on the complex manifold $\mathcal{R}(V) \subseteq V$; thus the given function f can be assumed to be bounded and holomorphic at least on $V-B$, where $B \subset V$ is the critical locus of the branched analytic covering. Now turning to the proof of Theorem 18, observe that in that proof it is really sufficient merely that the function f be bounded and holomorphic on $V-B$; for the coefficients of the various polynomials constructed in the proof of that theorem are then bounded and holomorphic in $U-D$, where $D = \pi(B)$ is a proper analytic subvariety of the domain $U \subseteq \mathbf{C}^k$, and hence by the usual Riemann removable singularities theorem once again they extend uniquely to holomorphic functions in U. With this observation made, the lemma is then an immediate consequence of Theorem 18.

This lemma can also be described in the following more convenient terms. A <u>weakly holomorphic function</u> in an open subset U of a complex analytic variety V is a function which is defined and holomorphic in $U \cap \mathcal{R}(V)$, and is bounded in the intersection of an open neighborhood of any point of U with the regular locus $\mathcal{R}(V)$; note that this local boundedness condition is non trivial only at points of the singular locus $\mathcal{S}(V)$. The weakly holomorphic functions in U clearly form a ring, which will be denoted by ${}_V\hat{\mathcal{O}}_U$; note that ${}_V\mathcal{O}_U \subseteq {}_V\hat{\mathcal{O}}_U$, with equality at least when $U \subseteq \mathcal{R}(V)$. The ring of germs of weakly holomorphic functions at a point $q \in V$ will be denoted by ${}_V\hat{\mathcal{O}}_q$; an element $f \in {}_V\hat{\mathcal{O}}_q$ of course need not be represented by a function which is well defined at the point q if $q \in \mathcal{S}(V)$. The collection of all the rings ${}_V\hat{\mathcal{O}}_q$ form a sheaf ${}_V\hat{\mathcal{O}}$ over the variety V; and $\Gamma(U, {}_V\hat{\mathcal{O}}) \cong {}_V\hat{\mathcal{O}}_U$, for any open subset $U \subseteq V$. Note that any bounded holomorphic function on the complement of a proper analytic subvariety of a connected analytic variety V is automatically a weakly holomorphic function in V.

Now for any germ $f \in {}_V\hat{\mathcal{O}}_q$ of a weakly holomorphic function on the pure dimensional analytic variety V at the point $q \in V$, it follows from the removable singularities lemma that

$f \in {}_V\mathcal{M}_q$ and that f is integral over the subring ${}_V\mathcal{O}_q \subset {}_V\mathcal{M}_q$. Conversely, if $f/g \in {}_V\mathcal{M}_q$ is integral over the subring ${}_V\mathcal{O}_q \subset {}_V\mathcal{M}_q$, then a representative function f/g is bounded and holomorphic in the complement of the analytic subvariety representing $\operatorname{loc} {}_V\mathcal{O}_q \cdot g$ in an open neighborhood of the point q; and hence $f/g \in {}_V\hat{\mathcal{O}}_q$. Thus the removable singularities lemma can be restated as follows.

Removable Singularities Lemma (b). For any point q of a pure dimensional complex analytic variety V, the ring ${}_V\hat{\mathcal{O}}_q$ of germs of weakly holomorphic functions is precisely the integral closure of the ring ${}_V\mathcal{O}_q$ in its total quotient ring ${}_V\mathcal{M}_q$.

With these preliminary observations and definitions out of the way, the result now of interest can be stated as follows.

Theorem 24. A germ $f \in {}_V\mathcal{M}_q$ of a meromorphic function at a point q of a pure dimensional analytic variety V is contained in the direct image $\rho_*({}_{V_1}\mathcal{O})_q$ under some germ of a simple analytic mapping $\rho: V_1 \longrightarrow V$ if and only if $f \in {}_V\hat{\mathcal{O}}_q$.

Proof. If $f \in {}_V\mathcal{M}_q$ is a meromorphic function such that $f \in \rho_*({}_{V_1}\mathcal{O})_q$ under a simple analytic mapping $\rho: V_1 \longrightarrow V_q$ over an open neighborhood V_q of the point q, then f is bounded in an open neighborhood of the point q, and hence $f \in {}_V\hat{\mathcal{O}}_q$. Conversely, consider a germ $f \in {}_V\hat{\mathcal{O}}_q$. Choose an open neighborhood V_q of the point q which can be represented as a branched analytic covering $\pi: V_q \longrightarrow U$ over an open domain $U \subseteq \mathbb{C}^k$. If this neighborhood is chosen sufficiently small, the variety V_q can be represented by a complex analytic subvariety V_q of an open subset $U \times U' \subseteq \mathbb{C}^k \times \mathbb{C}^{n-k} = \mathbb{C}^n$, with the point q corresponding to the origin, such that the mapping π is induced by the natural projection $U \times U' \longrightarrow U$; and the germ f can be represented by a weakly holomorphic function f on the variety V_q. The removable singularities lemma implies that there is a monic polynomial $p_f(X) \in {}_V\mathcal{O}_q[X]$ such that $p_f(f) = 0$ in ${}_V\mathcal{M}_q$. If the neighborhood V_q is chosen sufficiently small, this polynomial can be represented by a polynomial $p_f(X) \in {}_V\mathcal{O}_{V_q}[X]$ such that $p_f(f) = 0$ on V_q ; and moreover, there will exist a monic polynomial $P_f(X) \in {}_n\mathcal{O}_{U\times U'}[X]$ such that $P_f(X)|V_q = p_f(X)$. The removable singularities lemma also implies that $f \in {}_V\mathcal{M}_q$, hence that $f = h/g$ for some germs $g,h \in {}_V\mathcal{O}_q$ where g is not a zero divisor. Again, if the neighborhood V_q is chosen sufficiently small, these germs can be represented by holomorphic functions on V_q ; and moreover, there will exist holomorphic functions $G,H \in {}_n\mathcal{O}_{U\times U'}$ such that $G|V_q = g$ and $H|V_q = h$. Now consider the complex analytic variety

$$V_0 = \{(z_1,\ldots,z_{n+1}) \in U\times U'\times\mathbb{C} \,|\, (z_1,\ldots,z_n) \in V_q; P_f(z_{n+1}) = 0; G\cdot z_{n+1} - H = 0\}.$$

Note that the intersection $V_q \cap \{(z_1,\ldots,z_n)|z_1 = \ldots = z_k = 0\}$ is just the point $q = (0,\ldots,0) \in \mathbb{C}^n$, hence the intersection $V_0 \cap \{(z_1,\ldots,z_{n+1})|z_1 = \ldots = z_k = 0\}$ consists at most of the finitely many points $(0,\ldots,0,z_{n+1})$ for which z_{n+1} is one of the roots of the polynomial equation $P_f(z_{n+1}) = 0$; thus it follows from Theorem 9(b) that $\dim V_0 \leqq k = \dim V_q$, after shrinking the neighborhood V_q further if necessary. The natural projection mapping $U \times U' \times \mathbf{C} \longrightarrow U \times U'$ induces a complex analytic mapping ρ from the analytic subvariety $V_0 \subseteq U \times U' \times \mathbf{C}$ onto the analytic variety $V_q \subseteq U \times U'$. Since this projection mapping even induces a proper light mapping from the subvariety $\{(z_1, \ldots, z_{n+1}) \in U \times U' \times \mathbf{C} | P_f(z_{n+1}) = 0\}$ onto $U \times U'$, it follows that the restriction ρ of this mapping to the closed subvariety V_0 is also proper and light. Introducing the complex analytic subvarieties

$$A_0 = \{(z_1, \ldots, z_{n+1}) \in V_0 | G(z_1, \ldots, z_n) = 0\} \quad \text{and}$$

$$A_q = \{(z_1, \ldots, z_n) \in V_q | G(z_1, \ldots, z_n) = 0\} ,$$

note that $V_q - A_q$ is a dense open subset of V_q , since $g = G|V_q$ is not a zero divisor at any point of V_q and hence does not vanish identically on any component of V_q . The function $f = h/g$ is holomorphic on $V_q - A_q$, and $p_f(f) = 0$ at all points of $V_q - A_q$; consequently the mapping which associates to any point $(z_1, \ldots, z_n) \in V_q - A_q$ the point $(z_1, \ldots, z_n, f(z_1, \ldots, z_n)) \in U \times U' \times \mathbf{C}$ is an analytic mapping $F: V_q - A_q \longrightarrow V_0 - A_0$. This mapping F is clearly one-to-one, and has as its image the entire variety $V_0 - A_0$ since for any point $(z_1, \ldots, z_{n+1}) \in V_0 - A_0$ necessarily $z_{n+1} = H(z_1, \ldots, z_n)/G(z_1, \ldots, z_n) = f(z_1, \ldots, z_n)$; and the composition $\rho \circ F: V_q - A_q \longrightarrow V_q - A_q$ is the identity mapping, so that the restriction $\rho: V_0 - A_0 \longrightarrow V_q - A_q$ is an equivalence of complex analytic varieties. This is almost enough to show that $\rho: V_0 \longrightarrow V_q$ is a simple analytic mapping, except that it has not been verified that $V_0 - A_0$ is a dense open subset of V_0 ; but the latter assertion is not necessarily true. Thus it is necessary further to introduce the analytic subvariety $V_1 \subseteq V_0 \subset U \times U' \times \mathbf{C}$ consisting of those components of V_0 of pure dimension k on which the function G does not vanish identically, and the subvariety

$$A_1 = \{(z_1, \ldots, z_{n+1}) \in V_1 | G(z_1, \ldots, z_n) = 0\} .$$

The restriction of ρ to the subvariety $V_1 \subseteq V_0$ is still a proper, light, analytic mapping $\rho: V_1 \longrightarrow V_q$. Clearly $V_1 - A_1 \subseteq V_0 - A_0$; and since the variety $V_0 - A_0$ is of pure dimension k in an open neighborhood of each point, necessarily $V_0 - A_0 \subseteq V_1 - A_1$, so that indeed $V_1 - A_1 = V_0 - A_0$ and the restriction $\rho: V_1 - A_1 \longrightarrow V_q - A_q$ is hence an equivalence of complex analytic varieties. In this case $V_1 - A_1$ is evidently a dense open subset of V_1 , so that $\rho: V_1 \longrightarrow V_q$ is a simple analytic mapping. This mapping was so constructed that $\rho^*(f) = z_{n+1}|V_1 \in {}_{V_1}\mathcal{O}_{V_1}$, and hence $f \in \rho_*({}_{V_1}\mathcal{O})_q$; and that suffices to conclude the proof of the theorem.

The first consequences of this theorem are some simple additional properties of the weakly holomorphic functions on a complex analytic variety.

Corollary 1 to Theorem 24. If $d \in {}_V\mathcal{O}_q$ is a universal denominator on a pure dimensional complex analytic variety V, then $d \cdot {}_V\hat{\mathcal{O}}_q \subseteq {}_V\mathcal{O}_q$.

Proof. Given any germ $f \in {}_V\hat{\mathcal{O}}_q$, it follows from Theorem 24 that there exists a germ of a simple analytic mapping $\rho: V_1 \rightarrow V$ such that $f \in \rho_*({}_{V_1}\mathcal{O})_q$; hence $d \cdot f \in {}_V\mathcal{O}_q$, recalling the definition of a universal denominator.

Corollary 2 to Theorem 24. On a pure dimensional complex analytic variety V, ${}_V\hat{\mathcal{O}}_q$ is a finitely generated ${}_V\mathcal{O}_q$-submodule of ${}_V\mathcal{M}_q$.

Proof. Selecting a universal denominator $d \in {}_V\mathcal{O}_q$ which is not a zero divisor, and recalling the conclusion of Corollary 1, note that as ${}_V\mathcal{O}_q$-modules

$${}_V\hat{\mathcal{O}}_q \cong d \cdot {}_V\hat{\mathcal{O}}_q \subseteq {}_V\mathcal{O}_q \subseteq {}_V\mathcal{M}_q ;$$

that is to say, ${}_V\hat{\mathcal{O}}_q$ is isomorphic to the ideal $d \cdot {}_V\hat{\mathcal{O}}_q \subseteq {}_V\mathcal{O}_q$, hence is necessarily finitely generated.

One approach to the classification of simple analytic mappings over a pure dimensional complex analytic variety then follows from this next corollary and the corollary to Theorem 23.

Corollary 3 to Theorem 24. On a pure dimensional complex analytic variety V, the submodules $\rho_*({}_{V_1}\mathcal{O})_q \subset {}_V\mathcal{M}_q$ arising from germs of simple analytic mappings $\rho: V_1 \longrightarrow V$ are precisely the submodules of ${}_V\hat{\mathcal{O}}_q$ of the form ${}_V\mathcal{O}_q[f_1, \ldots, f_r]$ for some germs of weakly holomorphic functions $f_1, \ldots, f_r$ in ${}_V\hat{\mathcal{O}}_q$.

Proof. If $\rho\colon V_1 \longrightarrow V$ is the germ of a simple analytic mapping, it follows from Theorem 24 that $\rho_*({}_{V_1}\mathcal{O})_q$ is a submodule of ${}_V\hat{\mathcal{O}}_q$; this submodule is necessarily finitely generated, as a consequence of Corollary 2 to Theorem 24, and if $f_1,\ldots,f_r$ are module generators, then clearly $\rho_*({}_{V_1}\mathcal{O})_q = {}_V\mathcal{O}_q[f_1,\ldots,f_r]$. Conversely, consider a submodule ${}_V\mathcal{O}_q[f_1,\ldots,f_r] \subseteq {}_V\hat{\mathcal{O}}_q$. It follows from Theorem 24 that there exists a simple analytic mapping $\rho_1\colon V_1 \longrightarrow V$ such that $f_1 \in \rho_{1*}({}_{V_1}\mathcal{O})_q$; indeed, it is evident from the proof of that theorem that $\rho_{1*}({}_{V_1}\mathcal{O})_q \cong {}_V\mathcal{O}_q[f_1]$. (The simple analytic mapping $\rho_1\colon V_1 \longrightarrow V$ arises from a partial projection mapping, as in the local parametrization theorem; the function f_1 appears as the restriction of the coordinate z_{n+1} , and there is a monic polynomial in z_{n+1} in the ideal of the variety V_1 , so the argument is as on page 15.) The functions $f_2,\ldots,f_r$ induce weakly analytic functions $\overset{*}{\rho}_1(f_2),\ldots,\overset{*}{\rho}_1(f_r)$ on the variety V_1 . At each point of $\rho_1^{-1}(q) \subset V_1$ the argument can be repeated, using now the function $\overset{*}{\rho}_1(f_2)$; there results a simple analytic mapping $\rho_2\colon V_2 \longrightarrow V_1$, and the composite $\rho_1\rho_2\colon V_2 \longrightarrow V$ is a simple analytic mapping such that $(\rho_1\rho_2)_*({}_{V_2}\mathcal{O})_q \cong {}_V\mathcal{O}_q[f_1,f_2]$. The iteration of this argument then yields the proof of the desired result.

(e) There is another, more geometrical approach to the classification of germs of simple analytic mappings $\rho\colon V_1 \longrightarrow V$ over a pure dimensional complex analytic variety V at the point $q \in V$.

Since the ring ${}_V\hat{\mathcal{O}}_q$ of germs of weakly holomorphic functions is a finitely generated ${}_V\mathcal{O}_q$-module by Corollary 2 to Theorem 24, it follows from Corollary 3 to Theorem 24 that there exists a germ of a simple analytic mapping $\hat{\rho}\colon \hat{V} \longrightarrow V$ at the point $q \in V$ such that $\hat{\rho}_*({}_{\hat{V}}\mathcal{O})_q = {}_V\hat{\mathcal{O}}_q$; and it follows from the corollary to Theorem 23 that this simple analytic mapping is uniquely determined up to analytic equivalence. The simple analytic mapping $\hat{\rho}\colon \hat{V} \longrightarrow V$ will be called the <u>normalization</u> of the germ of the complex analytic variety V at the point $q \in V$. The germ of the variety V at the point q will be said to be <u>normal</u> if this normalization $\hat{\rho}\colon \hat{V} \longrightarrow V$ is an equivalence of analytic varieties; thus the germ of the pure dimensional variety V at the point q is normal precisely when ${}_V\hat{\mathcal{O}}_q = {}_V\mathcal{O}_q$, that is, when every germ of a weakly holomorphic function is holomorphic. The normal germs are just those germs of complex analytic varieties for which the Riemann removable singularities theorem holds in the same form as for complex manifolds. More algebraically, it follows from the Removable Singularities Lemma (b) that the germ of a pure dimensional complex analytic variety V at a point q is normal if and only if its local ring ${}_V\mathcal{O}_q$ is integrally closed in its total quotient ring. It is clear that if $\hat{\rho}\colon \hat{V} \longrightarrow V$ is the normalization of the germ V, then $\hat{V}$ is itself a normal analytic variety; for the simple analytic mapping $\hat{\rho}$ induces an isomorphism between the rings of weakly holomorphic functions on V and on $\hat{V}$.

The normalization $\hat{\rho}\colon \hat{V} \longrightarrow V$ is in a very natural sense the maximal simple analytic mapping over the pure dimensional variety V at the point $q \in V$. For if $\rho_1\colon V_1 \longrightarrow V$ is any germ of a simple analytic mapping over the variety V at the point $q \in V$, then of course $\rho_{1*}({}_{V_1}\mathcal{O})_q \subseteq {}_V\hat{\mathcal{O}}_q = \hat{\rho}_*({}_{\hat{V}}\mathcal{O})_q$; it follows from Theorem 23 that there exists a complex analytic mapping $F\colon \hat{V} \longrightarrow V_1$ such that $\rho_1 F = \hat{\rho}$, and it is clear that F is itself even a simple analytic mapping. Consequently all the simple analytic mappings over the variety V at the point q are necessarily factors of the normalization mapping; a geometrical approach to determining all the simple analytic mappings consists in finding the normalization and then examining the possible factorizations of the normalization. The problem is still a non trivial one in most concrete cases, but can be considered as somewhat better understood than before. No attempt will be made here to discuss this classification in further detail; but to round off the discussion, a few general properties of the normalization and of normal analytic varieties will be considered briefly.

It is evident that a normal germ of an analytic variety is irreducible; for if a germ V of an analytic variety is reducible, then $\mathcal{R}(V)$ has at least two connected components, and the function which is identically 0 on one component and identially 1 on the other components is weakly holomorphic but clearly not holomorphic. Thus the normalization $\hat{\rho}\colon \hat{V} \longrightarrow V$ involves at least the splitting apart of the separate components of the germ V ; the connected components of the germ $\hat{V}$ correspond to the irreducible components of the germ V. Actually somewhat more can be said, and will shortly be said.

Considering the normalization $\hat{\rho}: \hat{V} \longrightarrow V$ of a pure dimensional germ of analytic variety, represented as a simple analytic mapping $\hat{\rho}: \hat{V} \longrightarrow V$ between two complex analytic varieties, there are analytic subvarieties $A \subset V$ and $\hat{A} \subset \hat{V}$ such that the restriction $\hat{\rho}: \hat{V}-\hat{A} \longrightarrow V-A$ is an equivalence of complex analytic varieties; consequently there is a well defined analytic mapping $\varphi: V-A \longrightarrow \hat{V}-\hat{A}$ which is inverse to the mapping $\hat{\rho}: \hat{V}-\hat{A} \longrightarrow V-A$. Assuming that $\hat{V}$ is represented by a complex analytic subvariety $\hat{V}$ of an open subset of $\mathbb{C}^n$, the component functions of the mapping φ are evidently weakly holomorphic functions on V; and thus φ can be viewed as a weakly analytic mapping $\varphi: V \longrightarrow \hat{V}$ which is inverse to the mapping $\hat{\rho}: \hat{V} \longrightarrow V$. Of course the mapping φ is not necessarily a well defined mapping outside of the regular locus $\mathcal{R}(V) \subseteq V$; but in some cases it can be defined everywhere on V, as can be seen by use of the following auxiliary result.

Lemma. If f is a weakly holomorphic function on a complex analytic variety V and if V is irreducible at a point $q \in V$, then f extends uniquely to a continuous function on $\mathcal{R}(V) \cup q \subseteq V$.

Proof. If f is weakly holomorphic near q, then by the Removable Singularities Lemma there is a monic polynomial

$p_f(X) \in {}_V\mathcal{O}_{V_q}[X]$ in an open neighborhood V_q of q in V such that $p_f(f) = 0$ on $\mathcal{R}(V) \cap V_q$. If the distinct roots of the equation $p_f(X) = 0$ at the point q are $X_1,\ldots,X_r$, and if $U_1,\ldots,U_r$ are arbitrary disjoint open neighborhoods of these separate points in $\mathbf{C}$, then the roots of the equation $p_f(X) = 0$ will lie in the union $U_1 \cup \ldots \cup U_r$ at all points of V_q provided that V_q is chosen sufficiently small; for as is familiar, the roots of a monic polynomial are continuous functions of its coefficients. If V is irreducible at the point q, the neighborhood V_q can be so chosen that $V_q \cap \mathcal{R}(V)$ is connected; the values of the function f in $V_q \cap \mathcal{R}(V)$ must therefore be contained in a single neighborhood U_i, and defining $f(q) = X_i$ clearly yields the unique continuous extension of the function f to $(V_q \cap \mathcal{R}(V)) \cup q$. That suffices to conclude the proof.

Now if the analytic variety V is irreducible at every point, it follows from the preceding lemma that any weakly holomorphic function on V automatically extends to a continuous function on the entire point set V. In this case, then, there is a well defined continuous mapping $\varphi: V \longrightarrow \hat{V}$ which is weakly analytic and is inverse to the normalization; so the variety V and its normalization are then homeomorphic as topological spaces, differing only in that $\hat{V}$ has more holomorphic functions than has V, in the obvious sense. Of course, similar assertions hold for any simple analytic mapping $\rho_1: V_1 \longrightarrow V$, since as noted any such mapping is a factor of the normalization mapping $\hat{\rho}: \hat{V} \longrightarrow V$.

Even though no attempt will be made here to discuss normalization and normality in detail, one specific property really must in all conscience be mentioned, namely, that the set of points at at which a pure dimensional complex analytic variety is not normal form a proper analytic subvariety. This has a number of rather striking consequences and reformulations. For instance, this property is really equivalent, modulo results just established, to the property that the set of points at which a pure dimensional complex analytic variety is normal form an open set. If this latter condition holds, then in the normalization $\hat{\rho}: \hat{V} \longrightarrow V$ of the variety V at a point $q \in V$, the variety $\hat{V}$ will be normal at all points, after restricting V to an open neighborhood of the point q if necessary, and it then follows from Corollary 1 to Theorem 22 that the variety V is normal outside of a proper analytic subvariety; the converse is of course quite trivial. Another equivalent property is that the sheaf of germs of weakly holomorphic functions on a pure dimensional complex analytic variety is a coherent analytic sheaf. If this last condition holds, then the sheaf of germs of universal denominators has stalks $\mathcal{J}_q = \{f \in {}_V\mathcal{O}_q \mid f \cdot {}_V\hat{\mathcal{O}}_q \subseteq {}_V\mathcal{O}_q\}$, so is a coherent sheaf of ideals over V as in the argument at the end of the proof of Theorem 22(b); and the set of points at which the variety V is not normal is the locus of the sheaf of ideals $\mathcal{J}$, hence is a proper analytic subvariety of V. Conversely, assuming that the property holds as originally stated, then in a neighborhood of any point of the variety V the sheaf ${}_V\hat{\mathcal{O}}$ of weakly analytic functions coincides with the direct image sheaf $\hat{\rho}_*({}_{\hat{V}}\mathcal{O})$, hence ${}_V\hat{\mathcal{O}}$ is a coherent analytic sheaf as a consequence of Theorem 22(a). Yet another equivalent property is that the sheaf of germs of universal denominators is a coherent analytic sheaf. For if this last condition holds, the set of points of the analytic variety V at which the variety is not normal, which is the locus of the sheaf of ideals $\mathcal{d}$, is a proper analytic subvariety of V; conversely, assuming that the property holds as originally stated, in a neighborhood of any point of V the sheaf $\mathcal{d}$ of universal denominators coincides with the sheaf $\mathcal{d}(\hat{\rho})$ of relative denominators for the normalization, and hence is a coherent analytic sheaf as a consequence of Theorem 22(b). Finally, note that as a consequence of this property a pure dimensional complex analytic variety which is normal at a point q is normal and hence irreducible at all points in an open neighborhood of q; hence in the normalization $\hat{\rho}: \hat{V} \longrightarrow V$, the reducible branches of the variety V are separated at all points. It was noted earlier that V may be irreducible at a limit of points at which it is reducible, so that this splitting into irreducible branches is rather non trivial. The restriction to pure dimensional complex analytic varieties is not essential, but is merely a consequence of the fact that the present discussion of simple analytic mappings was limited to the case of pure dimensional complex analytic varieties for the sake of convenience. Noting that any complex analytic variety can be written as a union of pure dimensional components, and that the natural normalization of the entire variety is the disjoint union of the normalizations of these separate components, the extension of the discussion to analytic varieties which are not necessarily pure dimensional is quite obvious.

A beautifully simple and direct proof of this property, due to Grauert and Remmert, is as follows.

Theorem 25. The set of points at which a pure dimensional complex analytic variety is normal form an open subset.

Proof. Since the theorem is of a local character, there is no loss of generality in restricting attention to an open subset V of the variety for which there exists a holomorphic function d such that d is a universal denominator but not a zero divisor at each point of V. Introduce the analytic subvariety $W = \{z \in V | d(z) = 0\}$; and let $\mathcal{M}$ be the sheaf of ideals of this analytic subvariety, so that $\mathcal{M}$ is a coherent sheaf of ideals in the structure sheaf ${}_V\mathcal{O}$. For each point $z \in V$ the set $\mathcal{L}_z = \mathrm{Hom}_{{}_V\mathcal{O}_z}(\mathcal{M}_z, \mathcal{M}_z)$ of module homomorphisms from the ideal $\mathcal{M}_z$ into itself is a well defined module over the local ring ${}_V\mathcal{O}_z$; and the set of all of these modules form a coherent analytic sheaf $\mathcal{L}$ over the analytic variety V. (The proof of this assertion is straightforward, and will be left to the reader.) Note that to any germ $g \in {}_V\mathcal{O}_z$ there is naturally associated the homomorphism $\lambda_g \in \mathcal{L}_z$ defined by $\lambda_g(f) = gf$ for any $f \in \mathcal{M}_z$; this then establishes an inclusion ${}_V\mathcal{O}_z \subseteq \mathcal{L}_z$, which evidently corresponds to a sheaf inclusion ${}_V\mathcal{O} \subseteq \mathcal{L}$. Note further that for any elements $\lambda \in \mathcal{L}_z$ and $f \in \mathfrak{M}_z \subseteq {}_V\mathcal{O}_z$ for which f is not a zero divisor, the quotient $\lambda(f)/f$ is a well defined germ of a meromorphic function at the point z ; and that for any other element $g \in \mathfrak{M}_z \subseteq {}_V\mathcal{O}_z$, the product $g\cdot(\lambda(f)/f) = \lambda(fg)/f = \lambda(g) \in \mathfrak{M}_z \subseteq {}_V\mathcal{O}_z$, since λ is a module homomorphism. The meromorphic function $\lambda(f)/f$ thus has the property that $(\lambda(f)/f)\cdot \mathfrak{M}_z \subseteq \mathfrak{M}_z$, and consequently must be integral over the subring ${}_V\mathcal{O}_z \subset {}_V\mathcal{M}_z$, hence a weakly holomorphic function. It is apparent that the resulting germ $\lambda(f)/f \in {}_V\hat{\mathcal{O}}_z$ is independent of the choice of the germ $f \in \mathfrak{M}_z$, hence that there results a natural inclusion $\mathcal{L}_z \subseteq {}_V\hat{\mathcal{O}}_z$ corresponding to a sheaf inclusion $\mathcal{L} \subseteq {}_V\hat{\mathcal{O}}$. The resulting inclusion ${}_V\mathcal{O} \subseteq \mathcal{L} \subseteq {}_V\hat{\mathcal{O}}$ is clearly the natural inclusion of the holomorphic functions into the weakly holomorphic functions. To conclude the proof, it is only necessary to show that the variety V is normal at a point $z \in V$ precisely when ${}_V\mathcal{O}_z = \mathcal{L}_z$; for if V is normal at a point $q \in V$, so that ${}_V\mathcal{O}_q = \mathcal{L}_q$, that from the coherence of the sheaves ${}_V\mathcal{O}$ and $\mathcal{L}$ it follows that ${}_V\mathcal{O}_z = \mathcal{L}_z$ for all points z of an open neighborhood of q , hence that V is normal in an open neighborhood of q .

Now if V is normal at a point $z \in V$, then ${}_V\mathcal{O}_z = {}_V\hat{\mathcal{O}}_z$, so that necessarily ${}_V\mathcal{O}_z = \mathcal{L}_z$. Conversely, suppose that V is not normal at the point $z \in V$, so that ${}_V\mathcal{O}_z \subset {}_V\hat{\mathcal{O}}_z$. By the

Hilbert zero theorem, $\mathcal{M}_z = \text{id } W = \sqrt{{}_V\mathcal{O}_z \cdot d}$, hence $\mathcal{M}_z^\nu \subseteq {}_V\mathcal{O}_z \cdot d$ for some power ν; and since d is a universal denominator at the point z, necessarily $\mathcal{M}_z^\nu \cdot {}_V\hat{\mathcal{O}}_z \subseteq {}_V\mathcal{O}_z$. Choose the least integer ν for which the latter containment holds, noting that $\nu \geq 1$; thus $\mathcal{M}_z^\nu \cdot {}_V\hat{\mathcal{O}}_z \subseteq {}_V\mathcal{O}_z$ but $\mathcal{M}_z^{\nu-1} \cdot {}_V\hat{\mathcal{O}}_z \not\subseteq {}_V\mathcal{O}_z$, so that there is a weakly holomorphic function $g \in \mathcal{M}_z^{\nu-1} \cdot {}_V\hat{\mathcal{O}}_z$ for which $g \notin {}_V\mathcal{O}_z$. Note that for any germ $f \in \mathcal{M}_z$ it follows that $g \cdot f \in \mathcal{M}_z^{\nu-1} \cdot {}_V\hat{\mathcal{O}}_z \cdot \mathcal{M}_z \subseteq \mathcal{M}_z^\nu \cdot {}_V\hat{\mathcal{O}}_z \subseteq {}_V\mathcal{O}_z$; actually, since f vanishes on the subvariety $W \subset V$ and since g is everywhere bounded, the analytic function gf necessarily vanishes on W, hence $g \cdot f \in \mathcal{M}_z$. Thus multiplication by g is a homomorphism $\lambda \in \mathcal{L}_z$ such that $\lambda \notin {}_V\mathcal{O}_z \subseteq \mathcal{L}_z$, so that ${}_V\mathcal{O}_z \subset \mathcal{L}_z$. As noted earlier, that suffices to conclude the proof.

INDEX OF SYMBOLS

INDEX